Dialogues Concerning
Natural Numbers

PETER LANG
New York • Washington, D.C./Baltimore • Bern
Frankfurt am Main • Berlin • Brussels • Vienna • Oxford

Charles Sayward

Dialogues Concerning Natural Numbers

PETER LANG
New York • Washington, D.C./Baltimore • Bern
Frankfurt am Main • Berlin • Brussels • Vienna • Oxford

Library of Congress Cataloging-in-Publication Data

Sayward, Charles.
Dialogues concerning natural numbers / Charles Sayward.
p. cm.
Three of the dialogues have appeared journals in
2002 and 2005 and have been somewhat altered.
Includes bibliographical references and index.
1. Mathematics—Philosophy. I. Title.
QA8.4.S29   510.1—dc22   2009024860
ISBN 978-1-4331-0780-1

Bibliographic information published by **Die Deutsche Bibliothek**.
Die Deutsche Bibliothek lists this publication in the "Deutsche
Nationalbibliografie"; detailed bibliographic data is available
on the Internet at http://dnb.ddb.de/.

The paper in this book meets the guidelines for permanence and durability
of the Committee on Production Guidelines for Book Longevity
of the Council of Library Resources.

© 2009 Peter Lang Publishing, Inc., New York
29 Broadway, 18th floor, New York, NY 10006
www.peterlang.com

Printed in Germany

To Lela

Table of Contents

INTRODUCTION

The realist says that enormously many of the propositions accepted as premises for proofs in mathematics, along with the enormously many further propositions argued for in those proofs, assert the existence of numbers, and that these propositions are true. *Hence*, numbers exist— there really are such things.

The most common form of mathematical realism is mathematical Platonism. The Platonist says that numbers are not themselves objects with spatial locations or temporal durations. They are not themselves part of the physical, perceptually accessible world.

If the Platonist is right, numbers exist and are objects of a non-physical and perceptually inaccessible kind.

One response to the Platonist is *nominalism*. Nominalism agrees with Platonism in holding that the mathematical propositions we accept as premises for proofs in mathematics, along with the mathematical propositions argued for in those proofs, are true only if there are numbers, but differs from Platonism in holding that either those propositions are true but that the objects of which they speak are not things of a non-physical and perceptually inaccessible kind *or* that they are false since the objects of which they speak are indeed things of a non-physical and perceptually inaccessible kind and there are no such things.

The nominalist and Platonist conceptions of the character of mathematical propositions are essentially one. Where nominalism departs from realism is in holding that unless the objects of which these propositions speak are "nominalistically acceptable" these propositions are false.

What is a "nominalistically acceptable" object? Roughly this, any element or aspect of the physical world which is, directly or indirectly, perceptually accessible.

These two philosophical theories are the background of the six dialogues occurring herein.

There are five characters in these dialogues. Three of the characters, A, N, and C, are nominalists; while a fourth character, P, is a Platonist. The main character, M, appears in each dialogue. M is somewhat skeptical on most issues in the philosophy of mathematics. M is particularly skeptical regarding both background theories, nominalism and Platonism. This skepticism is most manifest in chapters 4 and 6.

M pushes a thesis made several years ago by Rudolph Carnap, who published an article that made this point: The sentence 'There are numbers' is a trivial truth if construed as an assertion from within mathematics. As such it cannot possibly divide philosophers of mathematics. It lacks sense if construed as an assertion outside of mathematics. (Carnap1978, p. 153) I do not think that the truth and importance of Carnap's thesis has been appreciated by philosophers of mathematics.

Three of the dialogues have appeared in print, somewhat altered. They are: (1) "Why Axiomatize Arithmetic?" *Sorites*, December 2005, Issue # 16: 54-61 (this appears as chapter 3 in this book); (2) "A Conversation about Numbers and Knowledge", *American Philosophical Quarterly*, July 2002, Volume 39, No.3: 275-287 (this appears as chapter 4 in this book); (3) "A Conversation about Numbers" (*Philosophia*, 2002, Volume 29, Nos. 1-4: 191-209 (this appears as chapter 5 in this book, entitled "A Conversation of Five"). I thank the editors of these journals for their permission to include these items in this book.

Philip Hugly has had a profound influence on my thinking in the philosophy of mathematics. I believe I attended every seminar he gave on the subject at Nebraska. The conversations we have had on the subject are replicated in this book.

CHAPTER 1
A Conversation on Frege and Mill

Scientific theories contain principles associated with observationally testable contingent statements whose truth or falsity respectively confirm or disconfirm those principles. Does the same hold for the propositions of mathematics? We may be tempted to say that such confirmation is irrelevant to, e.g., the elementary equations of the arithmetic of the natural numbers, since they are subject to *proof*. But *if* those equations were themselves associated with contingent and observationally testable statements whose truth or falsity respectively confirm or disconfirm those equations, then those equations would be confirmable or disconfirmable *apart from proof*, in which case the principles from which those equations ordinarily are proven would also be subject to confirmation and disconfirmation. The first dialogue discusses this issue.

The Dialogue

A. Mill felt that the elementary equations of arithmetic stated matters of fact (Mill 1974, p. 257).

M. But wasn't it generally agreed that these equations rested on definitions, in particular on the definitions of the numbers?

A. Yes, and Mill objected to this. He felt that the so-called definitions each asserted some matter of fact, and that this factual content carried over to all the equations.

M. So, where others might have said that '2 + 1 = 3' is a *definition* of 3, Mill held that this sentence asserted some matter of fact.

A. Exactly.

M. And what matter of fact did he think was asserted by this equation?

A. His idea was that it asserted that there are sensible objects that can be arranged into a single group of three, or separated into a single object and a pair of objects.

M. So his claim was that when we assert that 2 + 1 = 3 we are asserting that there are groups of two objects to which we can add one more object to form a group of three objects. So that if we *add* one object to a group of two objects the result is a group of three objects.

A. Yes. That was his idea.

M. And Frege said that *if* this were so, then if objects were all nailed down 2 + 1 would not equal 3 (Frege 1960, p. 9).

A. Yes. That's what he said. And he thought that this was a decisive objection to Mill's view. But in fact it wasn't. For Mill could always reply that it was *no objection at all* to his view, since his view *included* the idea that apart from our being able to arrange and rearrange objects the usual arithmetical equations would be false, not true.

  If it never happened that one object came into close spatial proximity with two other objects already in close spatial proximity, or if we could not make this happen, then it would *not* be the case that 2 + 1 = 3.

M. But doesn't that seem wrong? Imagine a field on which there are some marbles. Two marbles and just two are in contact. Exactly one marble is in contact with no marble. And there are no other marbles. Now, if things are as I have described them to be, how many marbles are there on that field?

A. Exactly three.

M. And wouldn't you say that the two marbles in contact with one another together with the marble in contact with no marble are three marbles altogether?

A. Of course.

M. And now suppose that I try to rearrange the marbles, but fail to do so. Will there then *not* be three marbles on the field?

A. Of course not.

M. Now suppose that more marbles get introduced onto the field, and that they move about, but that at no time do three marbles contact one another, and at no time are any three marbles at the same or even nearly the same distances from one another. Nor can we force them into such configurations. This then is a field in which there is no group of three marbles, and in which marbles cannot be arranged into groups of three. Do you agree?

A. Yes.

M. Now—would you want to say that *with respect to this field* it is *not* the case that 2 + 1 = 3?

A. Not really. For I can perfectly well see that there might be two marbles touching near a corner of the field and a third about an inch off from them and those two marbles and that one marble certainly are three marbles altogether.

M. And can't you now see that it just doesn't matter *how* the marbles are arranged or can be arranged. If there are exactly three marbles in some area, then there are two marbles in that area, and one more as well, and conversely.

A. Yes, I do really seem to see that.

M. So there really seems to be a problem in Mill's view, for he seems to have implied that two marbles and one more marble are three marbles *only if* they can be arranged in certain ways. But now we seem to see arrangement and rearrangement is quite unnecessary.

A. Was this what Frege had in mind when he said if Mill were right then if everything were nailed down 3 would not be 2 + 1?

M. I think so.

A. So there may have been *a* point in Frege's remark.

　　But not a *decisive* point, for it may be that Mill was wrong only about the kind of fact he thought the equation implied.

　　There must be something about them which enables us to separate and combine the marbles, at least *in thought*. So perhaps it is enough if spatial separation and combination are *conceivable*.

M. Here Frege would reply along the following lines: If Bill made exactly two promises to Tom and just one promise to Mary and no promises to anyone else, then he made three promises altogether.

But promises cannot even be *conceived* to be spatially separated and combined.

A. I should not have said *spatial*. All that is needed is that things fall into groups, whether spatial or non-spatial.

M. That sounds good enough for numbers greater than one, but Frege stressed that it seems odd for one and just impossible for zero. (Frege 1960, p. 9)

What would be a *group* of just one thing, or of no things at all?

A. Yes, I too can feel the awkwardness in speaking of a group of zero chairs, for example.

Still, as I reflect on it, I keep coming back to the idea that there is something right in Mill's ideas.

After all, don't we *find out* about sums by putting objects into groups in various ways and counting the objects in those groups or just seeing how many objects are in them?

I put some stones side by side on a table and see that they are two. Then I add one stone, and see that now there are three. This shows me that $2 + 1 = 3$.

So this *fact*—that adding one stone to a group of two stones produces a group of three stones—must at least *support* the claim that $2 + 1 = 3$.

And if we *never* ended up with three stones or three of anything else after such a process then surely we'd have no reason to say that $2 + 1 = 3$.

And so the equation that $2 + 1$ equals 3 really does seem to be connected with matters of fact.

M. Do you think that learning to add is *finding out* what the sums are?

A. Of course. For what you learn is what you find out, and in learning to add you find out such things as that $2 + 1$ equals 3.

Take a look at a simple case of adding. You form two groups of marbles, count the marbles in each group, put the marbles together, and then count them all. So you find out that when 6 marbles are combined with 5 other marbles the result is 11 marbles.

M. But surely the combining of the marbles is inessential. It would be enough to count first the one group, then the second group, and then count through both groups. You would have found out the result of adding 5 to 6 *without* any rearrangement of the marbles.

A. Fair enough. And now I remember Frege saying that *adding* is not *combining*. I can see what he means.

But still there is a *fact* here—that if there are 5 marbles in one group of marbles and 6 marbles in another group of marbles, then there will be 11 marbles in the two groups altogether.

M. But now suppose that there is a time at which there are 5 marbles in one group and 6 marbles in the other group. And then, a moment later, there are 10 marbles in the two groups altogether.

Would this be a fact of a kind that went against 5 + 6 being 11?

A. Certainly.

M. And what about this: There is a time at which there are 5 marbles in one group and 6 marbles in the other group and 10 marbles in the two groups altogether—*at the very same time.*

Would this be a fact of a kind that went against 5 + 6 being 11?

A. Sure.

M. And what about this: Suppose there is just one red marble on the table and just one blue marble on the table and no other marbles on the table and no marbles on the table which are both red and blue *and* just one marble on the table.

Would this be a fact of a kind that went against 1 + 1 being 2?

A. I see that to be consistent with my earlier reply I have to say 'Yes' to this. But to be honest, it seems to me that there could be no such fact.

It couldn't be that there was one marble on table and one more marble on the table without there being two marbles on the table.

And as I reflect on it, I think I can see that there couldn't be two groups of marbles with 10 marbles altogether *and* with 6 in one group and 5 in the other group and none in both.

M. Should we then conclude that nothing could go against 6 + 5 equaling 11?

A. I don't think so. We surely could have evidence that 6 + 5 does not equal 11.

M. What might be such evidence?

A. Well, we count the two groups and then count the marbles altogether and take the second count as providing evidence for how many marbles there were when the first counts took place.

   So, there can be evidence—even very good evidence—that at the time at which the two groups contained 5 and 6 marbles respectively, with no marble in common, they contained 10 marbles altogether.

M. If I understand you, the evidence would be as follows: We get to 5 in counting the marbles in the first group, and to 6 in counting the marbles in the second group, and to 10 in counting the marbles in the two groups.

   So we have evidence for *these* conclusions: At the time of the first counting there were 5 marbles in the first group, and at the time of the second counting there were 6 marbles in the second group, and at the time of the third counting there were 10 marbles in the two groups altogether.

   How is that good evidence that 6 + 5 equals 10, or that 6 + 5 does not equal 11?

A. Well, you left out one point, that the evidence that there were 10 marbles at the time of the overall counting is *also* evidence that there were 10 marbles earlier on.

M. So, the idea is that you know already that in general if there were 10 stones at one time, then there were 10 stones just before that time.

A. Yes, experience teaches us that.

M. But does experience teach us that *for this case*?

   We both see that *if* there were 6 stones in one group and 5 stones in the other group at some time, then *at that time* there were altogether 11 stones in the two groups. And so we know that *if* there were altogether just 10 stones a moment later, then there was *one stone less*. And we can speculate about what happened to whichever stone it was which wasn't there a moment later.

A. But *must* we see that if there were 6 stones in one group and 5 in another group then *at that time* there were altogether 11 stones?

Couldn't the course of our experience be such as to finally give us *very good grounds* for concluding that two such groups comprise 10 and not 11 stones?

M. Perhaps.

A. So, you admit that it is *possible*. But then our experience *could* give us good grounds for thinking that 6 + 5 equals 10 and not 11.

M. That's too fast. I said 'perhaps', not 'possibly'. I was only indicating that I could not rule this out right off.

I admitted that *for all I can now show* experience could give us good grounds for thinking that 6 + 5 equals 10 and not 11. But that is not the same as saying that it is *possible* that experience *could* give us such grounds.

A. Still, you admit that for all you *know* experience could give us good grounds for accepting equations other than those we currently accept.

So you at least don't *know* that Mill was wrong.

M. It is likely that I *don't* know that Mill was wrong. These issues are difficult to deal with, and I don't as yet have a clear view of them.

But what Mill seems to have held is not that experience can give us grounds either for accepting or rejecting the usual equations, but that those equations assert matters of fact, assert facts which could have been otherwise.

But you agree that it *couldn't* be the case that two groups of 5 and 6 marbles respectively, with no marble in common, together comprise anything other than 11 marbles.

So, what our equation asserts about groups of marbles, or whatever, *can't* be otherwise.

And so it seems that Mill was wrong.

A. I think you are right on this, so far as it goes. I think that Mill *did* think that our equations express matters of *contingent* fact. And I agree that the possibility of having good *grounds* for rejecting these equations does not show that what the equations assert are contingent facts.

After all, for all I know it is possible for there to be situations in which we have good grounds for rejecting necessary truths!

M. So, then, it seems that Mill really was in error.

A. I'm not ready to draw that conclusion. There are *two* facts here. One is that two groups of 5 and 6 marbles respectively, with no marble in common, together comprise 11 marbles. The other is that if two groups of 5 and 6 marbles respectively, with no marble in common, were combined there would be exactly 11 marbles altogether. The second fact, unlike the first, is contingent.

Couldn't Mill insist that our equation asserts the second fact, the contingent one?

M. I suppose he could. I am not sure I would want to argue the point with him. I might instead propose that we use two forms of speech for the two cases. We could say either that 5 and 6 *are* 11 or that 5 and 6 *make* 11. The first equation would hold just in case, for any groups A and B, if at some one time there are 5 objects in A and 6 objects in B and no object in both A and B, then there are 11 objects in A and B altogether at that time. The other equation holds only if, in addition, there are still 11 objects in A and B altogether a moment later.

So, the assertion that 5 and 6 *are* 11 does not assert anything that could fail to be the case, but the assertion that 5 and 6 *make* 11 does.

We can continue to say that the assertion that 5 and 6 are 11 implies facts, but not facts of a kind which could fail to obtain. And this is what Mill had in mind—matters of fact *that could be otherwise.*

A. I see what you mean.

M. Now, I think that Frege took the ordinary '5 + 6 = 11' to be saying simply that 5 and 6 *are* 11, so that no facts which *could* be otherwise were entailed—and thus no facts about separating and bringing together. For such facts *could* be otherwise.

A. So *that* was what he was driving at when he spoke of 'nailing down' objects!

M. I think so. His basic point need not take the form of denying that mathematical equations assert facts, but of denying that they assert facts which could be otherwise in the way that ordinary physical facts—the sorts of facts Mill had in mind—could be otherwise.

A. But might he not have been wrong about our ordinary equations—that they do not say that 5 + 6 *are* 11 but that 5 + 6 *make* 11?

M. I suppose so. But that really isn't important. What is important is whether or not there are mathematical *necessities*. And we have both seemed to see that there are.

A. Perhaps, but I still come back to the fact that we teach addition by arranging things, putting them together, and separating them.

M. We do.

A. And isn't this a matter of giving the learner examples? So that by adding a third marble to a group of two we bring it about that two of *those* marbles plus one more of *those* marbles equals three marbles?

M. Well—we teach the child to add. And it proves useful in doing this to spatially arrange and rearrange objects such as marbles. But we might place three marbles on the table, two being red and one being green, and say 'And those two red marbles plus this one green one make three marbles altogether'. Here there is no rearrangement *at all*, but the technique works in teaching the child how to add.

A. But we *can* do it the other way.

M. Of course. But whether we teach it with one technique or the other we teach the same thing: sums.

　　Or do you think that with one technique we teach one type of sum, the kind which entails certain matters of contingent fact, whereas with the other technique we teach another type of sum, a kind which entails no matters of contingent fact?

A. No—it really does seem to me that we teach just the same thing either way.

M. So the sums we teach—for example, that 5 + 6 equals 11—do *not* imply any matters of contingent fact. So that Mill was wrong on this point.

A. Perhaps. But it seems to me *very* strange that our sums should state no facts at all.

M. Frege might have said that they state *mathematical* facts.

A. Yes—mathematical facts *known in some special way*. And that is the worrisome part.

M. Suppose we inspect an animal taken from the sea and conclude it is a whale. That is an experiential judgment. Suppose someone else inspects that animal and concludes it is not a whale. That too is an experiential judgment. Do we now have evidence that the animal under inspection is both a whale and not a whale?

A. Of course not. We now have evidence that someone has drawn a false conclusion from their inspection of that animal.

M. So the inference from 'There is evidence that p' and 'There is evidence that q' to 'There is evidence that p and q' falls short of validity.

A. But that is not to say that the inference is always in error. If there is evidence that it snowed in Nebraska in the winter of 1876, and there is evidence that the temperature fell below the freezing point in Nebraska in the winter of 1876, then there is evidence that it both snowed and got below zero in Nebraska that winter.

M. Here logic serves as a criterion for the combining of evidence. If p and q are logically inconsistent then evidence for p and evidence for q do not combine to form evidence for p and q.

A. Right. But this is not to say that we may not *believe* that our evidence that p and our evidence that q jointly form evidence that p and q even when p and q are logically inconsistent. But, I agree, if p and q are logically inconsistent, then the evidence that p together with the evidence that q is not evidence that p and q.

M. Might it not be like this in mathematics?

A. How so?

M. Well, imagine this situation. There is evidence that the apples to the left are 4 and there is evidence that the apples to the right are 5 and there is evidence that the apples to the right and left together are 8, but there is no evidence that 4 apples to the left together with 5 apples to the right are 8 apples altogether.

It may be that like logic, mathematics serves as a criterion for the combining of evidence. If $n + m = k$ is mathematically inconsistent then evidence that there are n apples to the left and evidence that there are m apples to the right and evidence that there are altogether k apples to the left and the right do not combine to form evidence that n apples to the left and m apples to the right are k apples altogether

A. So, this is the situation. I count the apples to the left and conclude the apples to the left are 5. I do the same for the apples to the right and conclude that they are 4. Then I conclude that there are 5 + 4 apples altogether. I also count all the apples and conclude that there are 8 apples altogether.

Then the question then becomes: Do I then have evidence that 5 + 4 apples are 8 apples?

M. And the answer, I guess, is this: I now have evidence that I have drawn a false conclusion from my apple countings.

CHAPTER 2
A Conversation about the Peano Axioms

The arithmetic of the system of natural numbers, set out as a first-order theory, is sometimes called 'elementary number theory', sometimes ' number theory', sometimes 'arithmetic', and sometimes (when developed in terms of the Peano axioms) 'Peano arithmetic'. The term 'arithmetic' will be favored here.

The Peano axioms for the natural numbers are formulated in a first order language of the usual type (connectives, an identity predicate, individual variables, and quantifiers) together with, for example, the following non-logical vocabulary: the unary operation symbol 's', the binary operation symbols '+' and '·', and the individual constant '0'. The latter symbols can be given their usual arithmetical readings. The first symbol can be read as 'the immediate successor of'.

Here they are:

1. $\forall n \, (0 \neq s \, (n))$

2. $\forall n \, \forall m \, (s \, (n) = s \, (m) \supset n = m)$

3. $\forall n \, (n + 0 = n)$

4. $\forall n \, \forall m \, (n + s \, (m) = s \, (n + m))$

5. $\forall n \, (n \cdot 0 = 0)$

6. $\forall n \, \forall m \, (n \cdot s \, (m) = (n \cdot m) + n)$

7. All sentences of the form: $(A \, (0) \wedge \forall v \, (A \, (v) \supset A \, (s \, (v)))) \supset \forall v \, A \, (v)$

The conversation progresses to two distinctions: (1) a distinction between quantificational arithmetic and non-quantificational arithmetic (2) a distinction between deduction and calculation. The dialogue ends with an argument for mathematical realism.

In this dialogue M is joined by C. C begins with a question about understanding the axioms

The Dialogue

C. Let's talk about the axioms for arithmetic. They seem very simple, and yet I always feel less than entirely clear about them. I don't mean that I can't look them up or carry out deductions from them. But I somehow am not sure what connection they have with arithmetic. Certainly, arithmetic never *feels* like deduction when I actually do it.

   I just feel fuzzy on the whole topic.

M. OK. Then tell me what you think are the *very simplest* things in arithmetic—the things you would have first learned as a child.

C. That's easy to answer. You first learn some of the number words, and then to count with them, and then to add, multiply, subtract and divide.

M. That sounds about right. And now think for a moment about subtraction. If you subtract 3 from 5 you get 2. But what if you subtract 5 from 3?

C. Then you get minus 2, or negative 2.

M. Right. And if you divide 4 by 2, but if you reverse it and divide 2 by 4 you get a fraction, 1/2.

   But both1/2 and –2, the fractions and the negative numbers, are not simple counting numbers. Isn't that so? And didn't it take some special learning to get the hang of the negative numbers and the fractions?

C. Certainly.

M. So if we stick to the *simplest* parts of arithmetic we'll have to drop subtraction and division. And that leaves us just with the number words, counting, adding, and multiplying. Right?

C. Yes. That's clear.

M. OK. Then let's start with the number words and counting, since they actually come first. What do we use counting for? I mean, why do we count?

C. Well, to find out the numbers of things. If you want to know how many apples are in the basket, then unless there are just a few, so you can tell by looking, you need to count them out.

M. So it seems. We count to find out how many things of this or that sort there are. And we say how many things there are by giving a number. So counting is a way of getting to the numbers of things.

C. Yes.

M. OK. Now, let's leave counting aside for a moment and reflect just on the number of things. Tell me, what is the smallest number of things there might be?

C. One, I suppose. Though you don't have to count in that case.

M. Yes. You wouldn't have to count if there's just one. But just concentrate on the number, however you arrive at it. Is one the smallest number of things? Suppose I asked you for the number of mountains in Nebraska. What would you answer?

C. I'd say there aren't any at all.

M. But could you give that answer by a number?

C. You mean zero? Yes. I suppose so. That would be the number I'd give. I'd say that the number of mountains in Nebraska is zero.

M. And what would be the next smallest number? Any child knows the answer. One, of course. And next would come two, and then three. And, if you agree to leave out the fractions, and stick just with the numbers which say how many things there are, there are no numbers between zero and one, or one and two, or two and three. And as you go from one number to the next you always get to a new number. There will be no one number that comes after both of two different numbers. So the series of numbers never cycles back and starts over. I know it is almost ridiculous to make such obvious remarks, but they are correct aren't they?

C. They certainly seem to be correct. And I suppose it is just such simple things as these that the axioms say?

M. Yes. The first of the usual axioms says that zero is the smallest number among the numbers that give the numbers of things—the so-called natural numbers—and the second says that these numbers never circle back. These are the first two axioms. And they say the ridiculously simple and obvious things we've just been saying.

One says that zero does not come directly after any natural number, which is just a way of saying that zero is the smallest natural number. And the other says that no natural number comes directly after two different natural numbers, which is just a way of saying that as you go from one natural number to the next you never go back to an earlier natural number.

C. And the phrase 'immediate successor' that is there in the axioms, I suppose it has to do with coming directly after, with there being no natural number in between. Is that right?

M. Yes. That phrase, with that meaning, is what is there in the symbol '*s*' that is used in writing out the axioms. So if you write '*s* (0)' what is meant is the number which comes right after zero, and if you write '*s* (*s* (0))' what is meant is the number which comes right after that one. And so forth.

And so you can write the axioms this way:

For any number n, $0 \neq s(n)$

For any numbers n and m, if $n \neq m$, then $s(n) \neq s(m)$

As for the numbers 1, 2, 3, and the rest, well 1 is the successor of zero, and 2 is the successor of the successor of zero, and so forth. In general, you have the following correspondence:

$s(0)$	1
$s(s(0))$	2
$s(s(s(0)))$	3

and so on. To get the usual numeral from the notation using the letter '*s*', just count up the number of its occurrences and use the numeral you thereby arrive at.

C. Yes. That is pretty obvious.

M. So here we have just *two* axioms along with the numerals formed with '0' and '*s*'. These axioms, by the very simplest one step deductions, yield their instances, from which we can then reason to obtain a great many equalities and inequalities formed just with those numerals, the identity sign, and the slash for negation.

C. That does seem clear to me, though it still somehow feels odd to say that we *reason* to conclusions like $s(s(0)) \neq s(s(s(s(s(0)))))$. But I do have a feel for what is going on now, with the first two axioms at any rate. What about the others?

M. Well—it is helpful to get clear as to what you want from the other axioms. Given a pair of numbers, whether the same or different, you certainly would like to know its sum. Right?

C. Yes, for the simple cases of $n + m = k$, that's just what we're looking for.

M. And which are the simple cases? Aren't they those in which the equation is written with *numerals* in for n, m and k?

C. Yes. And since we're writing the numerals with '0' and 's', the numerals will be just the symbol for zero and the successor symbols built up from the symbol for zero by putting the *successor* sign in front of it. Further, I think I see that what we want is some method for determining, given a pair of numbers written with numerals, what their sum is, *also* written with a numeral.

M. I agree. Now won't the simplest case be that of adding zero? For in that case you just get what you began with. Adding zero makes no difference. I suppose that is very clear to you.

C. Of course. And I can see that the first axiom for addition, for every n, $n + 0 = n$, says just this. But it has always been the second axiom for addition that bothered me. I have heard it described as a definition. But that seems wrong, since the sign to be defined would be the '+' and it actually is *used* in the second axiom for addition.

M. Don't worry about whether or not to speak of *definition* in connection with the axioms for addition. Let us just understand the axiom.

To get the idea lying behind it, suppose you had a pair of numerals and already knew which numeral gave the sum of the numbers written with those numerals. Let the two numerals be J and I, and let the numeral that gives the sum of the numbers they represent be K. Now consider the sum of the numbers given by the numerals J and *s* I—that is, the numeral which results from putting an 's' in front of the numeral I. What will be the numeral that gives the sum in this case?

C. It will be the numeral that results from putting an 's' in front of the numeral K.

M. OK. That means that if you have already worked out the equation J + I = K then you can work out J + *s* I. You simply take the numeral K and put an 's' in front of it. Do you see this?

C. Certainly.

M. Then you understand the second axiom for addition. For one way of writing that axiom is this:

For every *n*, *m* and *k*, if $n + m = k$ then $n + s(m) = s(k)$

Given this, and the first axiom, any sum can be settled just by working up to it from simpler sums.

C. How so?

M. Well, suppose that the numerals J and I have no occurrences of 's'. Then our only question is this:

$$0 + 0 = ?$$

By the first axiom we have the answer, namely that $0 + 0 = 0$.

   Now suppose that the numerals J and I have together just one occurrence of 's'. Then we have just these questions:

$$s(0) + 0 = ?$$
$$0 + s(0) = ?$$

By the first axiom we have an answer to the first question, for by that axiom $s(0) + 0 = s(0)$. By the second axiom we know that if $0 + 0 = k$, then $0 + s(0) = s(k)$. Since we have already determined that $0 + 0 = 0$, we know that for this case $k = 0$. Thus, $s(k) = s(0)$ and we have an answer to the second question, namely that $0 + s(0) = s(0)$. Now suppose that there are two occurrences of 's' in J and I together. We then have these questions:

$$s(s(0)) + 0 = ?$$
$$0 + s(s(0)) = ?$$
$$s(0) + s(0) = ?$$

Well, the first axiom answers the first question. By the second axiom we know that if $0 + s(0) = k$, then $s(0) + s(s(0)) = s(k)$. But we have already determined that $s(0) + 0 = s(0)$. So $k = s(0)$. In that case $s(k) = s(s(0))$. And so we have an answer to the second question, namely that $0 + s(s(0)) = s(s(0))$. Next, we know from the second axiom that if $s(0) + 0 = k$, then $s(0) + s(0) = s(k)$. But we have already determined that $s(0) + 0 = s(0)$. So, $k = s(0)$, and $s(k) = s(s(0))$. This gives us the answer to the third question, namely that $s(0) + s(0) = s(s(0))$.

In this step by step way we could settle any sum by applying the two axioms for addition and making use of results of applying those axioms to simpler cases.

C. I think I see that.

M. And if you have a more complex addition, say something of the form $n + m = u + w$, well, you first figure out $n + m = z$ and then $u + w = t$, and if the same numeral gives both t and z, then you know that $n + m = u + w$. And if not, then you know that $n + m \neq u + w$. This should give you a feel for how you could figure out any addition, simple or complex, just by using the two axioms.

C. It does. And I can see roughly how the same will work for multiplication. But one thing about multiplication isn't clear to me. In the case of addition, the second axiom uses the plus sign. But the second axiom for multiplication goes back to the plus sign and doesn't use the times sign. Why is this? Why does *addition* come into multiplication when multiplication doesn't come into addition? In that respect the second axioms for addition and multiplication are quite different.

M. Well—it has to do with what multiplication is. Imagine a case in which you'd multiply. Maybe you count five tables in the room and six people at each table. Then you'd multiply five times six to get thirty altogether. You wouldn't bother to count out all the people one by one. But you could also just add six to itself five times. This shows what multiplication comes to. When you say 'Five times six' all that means is six plus six plus six plus six plus six—that is, the resulting of adding six to itself five times.

And this makes it clear why any number times zero is zero—for no matter how often you add zero to itself, you just get zero.

C. What then of zero times some other number. Say three. That would have to be, by your account, a matter of adding three to itself zero times. But what is that?

M   I just don't know. My *suspicion* is that it means nothing at all, and that we write zero times three equals zero just to keep multiplication like addition—that is, indifferent to order. But I have never thought the matter through.

C. Still, what you said about multiplication and addition is helpful. It is strange how I sometimes fail to see such simple things. I almost feel ashamed for having asked about something so evident.

M. These simple and familiar points probably are evident to lots of people. But often I need to think about these things to get clear about them. For example, I would not be able to say why you *cannot* define multiplication in terms of itself as you can with addition. That is something I have not thought through. And I have actually thought about why it should be that a negative number times a negative number is always a positive number, but without as yet getting clear about it.

C. Then I won't feel shy about asking questions when things are not clear to me, at least in our conversations.

But now, what now about the so-called induction axiom?

M. There isn't just *one* induction axiom, but lots of them. And they are very special. Their primary role is quite different from the primary role of the other axioms.

To help get clear about the induction axioms, let's go back a bit. I want to give you another way of axiomatically dealing with the equalities and inequalities deducible from the first two axioms. We will still work with the ideas that zero is the smallest number and that the numbers don't circle back. But this time we go directly to the instances of the axioms, and let them be the axioms.

C. What do you mean?

M. Well, consider *all* the inequalities of the form

$$0 \neq S(0)$$

where what goes in for the capital 'S' is a string of one or more occurrences of the small 's'. You could right off call *each* of these inequalities an axiom. If you did that you would then call

$$0 \neq S(0)$$

an axiom *schemata*, not an axiom. The *axioms* would be all the formulas you can get from the schemata by replacing upper case 'S' by a string of one or more occurrences of lower case 's'. Do you see how this works?

C. I think so.

M. Good. And then you could consider the schemata

$$\text{If } J \neq I, \text{ then } s J \neq s I$$

and take as axioms all the instances of this schemata which result from replacing 'J' and 'I' by various instances of the form 'S0'. Then you'd have as *axioms* such conditionals as

$$\text{If } s(0) \neq s(s(s(0))), \text{ then } s(s(0)) \neq s(s(s(s(0)))).$$

C. I see this. But why use these schemata? Why not just stick with the usual axioms and infer these 'axioms' as consequences? Introducing axiom schemata and defining the axioms as their instances seems only to make everything more complicated.

M. Well—the induction axioms are given by schemata. So the type of procedure I've just introduced is one we'll need to understand when we deal with induction.

C. OK.

M. But now that I look at it, it strikes me that the usual axioms are formulated precisely to provide for all the particular inequalities and conditionals which we've just specified as axioms. For all these particular inequalities and conditionals—the instances of the axiom schemata—are directly derivable from one or the other of the two axioms, and all the deductions start from there. So, it now seems to me that the axiom schemata are just another and equally good way of specifying those inequalities and conditionals.

And, as I reflect, it seems to me right to say that they really are all as much axiomatic as one another?

C. I don't understand that. I thought the *axioms* are axiomatic, and that we use them to establish the inequalities and the conditionals.

M. But are the usual axioms any more axiomatic than their instances?

We have, in everyday life, a method for writing out the numbers. We call those symbols the *numerals*. Part of what we learn in learning them is when they are different. We don't say of two marks that they are different numerals because of just any difference between them. Difference of color makes no difference, and difference of size makes no difference. But there are the familiar differences that make a difference. And these distinguish the numerals. Basically, we distinguish ten different shapes, and then say that two strings made up of figures of those shapes are the same numeral only if they are, in terms of those shapes, just the same. Then—if the numerals are the *same*, we put a '=' between them, but if they are different we put a '≠' between them. Don't the first two axioms sum this up?

C. Well—they lead to these results. For by means of those two axioms we can *prove* all these equalities and inequalities.

M. But isn't it wrong to say that it is by these *axioms* that we prove the *equalities*? If we think of proof here, won't it be simply a matter of the law of identity—that n is n?

C. Yes. The first two axioms are needed only to prove the inequalities. With them it's a matter of *deduction*: we use *logic* to *infer* the inequalities from these axioms. Unlike the equalities, we can't just write them down as truths of logic.

M. Agreed. We can carry out deductions here. But *need* we do so? And if we do carry out a deduction to arrive at the conclusion that $s\,(s\,(0)) \neq s\,(s\,(s\,(s\,(s\,(0)))))$, does that *show* us something new? So that we can be said to have *found out* that $s\,(s\,(0)) \neq s\,(s\,(s\,(s\,(s\,(0)))))$ by first knowing that zero is the immediate successor of no number and that different numbers have different immediate successors, and then *reasoning* from that to the new piece of information?

C. Are you asking whether, for example, we might know that zero is the immediate successor of no number and not as yet know that zero is not the successor of zero?

M. Yes. And I think I know that zero is not its immediate successor as soon as I bring it to mind. Or I might put it this way: As soon as I look at

$$0 \neq s\,(0)$$

I see that it is correct. If I didn't already know that, I wouldn't know that zero is the immediate successor of no number.

C. I feel the force of saying that. But what then of our knowledge that zero is the immediate successor of no number? I can't infer that from each of its infinitely many instances. For I can't know each of those, precisely because there are that many. So—won't my knowledge that zero is the immediate successor of no number come *before* my knowledge of infinitely many of its instances, even if not before my knowledge of all of them?

M. I'm not sure what to say. The relation between the axioms what we deduce from them is very unclear to me.

I sometimes feel that we begin with knowing that what the axioms say is so, and then by proof come to know the things those axioms entail. But then I feel that I already know lots of the things we might infer from the axioms. And then I ask whether I know the axioms on the basis of knowing these other things, and I feel I can't, since that would be inferring statements about all of the infinitely many numbers there are from some finite sample. And I hardly can see how we could get knowledge in that way!

C. Yes. It always amazes me how what at first seems so clear to me becomes unclear as soon as I start thinking about it. It's almost as if it takes *not* thinking to be clear about things!

M. I think there is something to that. But as you spoke it suddenly struck me that you could take all of the equalities and inequalities that are deducible from the first two axioms and make every last one of them an axiom.

And this even seems right to me, for I really cannot see that any one of them is any more basic than any other.

C. But how could they all be axioms?

M. Well, we'd specify them all as axioms by the kind of statement I've just made. Or, to put it in more professional style, you could use just one axiom schemata:

$$J \neq I$$

where what goes in for 'J' and 'I' are successor symbols ending with '0' but with different numbers of occurrences of 's' — counting a '0' all by itself as the successor symbol ending with '0' and prefixed by zero occurrences of 's'. And if the number of occurrences of 's' in J and I are the same, then you write

$$J = I$$

and that's the end of it!

C. So, you're saying that we could bypass all the usual proof, and even the appeal to the law of identity, and just lay down two axiom schemata,

$$J \neq I$$
$$J = J$$

with all their instances as axioms. You then have all the equalities and inequalities *without any reasoning* and even *without any appeal to logic.*

Is that what you're driving at?

M. I wouldn't say that is what I'm *driving at*—for I find myself saying things I in no way anticipated saying. But it surely does appear that all the equalities and inequalities can be specified independently of logic and deduction.

C. Yes. Without any deductions! That certainly fits in with my sense that somehow I'm not doing any deductive reasoning in doing arithmetic. But I'm not sure we should say, that no reasoning of any kind is involved. For it strikes me that though we're trying to get clear about arithmetic, the way we're now talking about the numerals itself requires some arithmetical thinking.

I mean, to see whether or not a given formula is an instance of one of these axiom schemata you need to *count* the occurrences of 's' to the right and left of the identity sign and see whether there are the same or different numbers of them.

And now it suddenly seems to me that the way you've suggested for deciding the equalities and inequalities, though it works without invoking logical inference, calls upon the very thing it seeks to decide—the recognition of same and different among the numbers!

That somehow makes the whole thing feel circular.

M. The way I put it does make it seem that we need to recognize that the number of occurrences of 's' to the right *is* or *isn't* the number of occurrences of 's' to the left. But it isn't necessary to actually do any counting or bring in numbers at all. All you need do is see whether there are *as many* occurrences of 's' on one side as on the other. For that you don't have to count or do anything else to find out the number of occurrences. It is enough to *correlate* them one by one. If when you've correlated all of them on one side there are none left on the other side, well—you know you have an instance of the second axiom schemata. Otherwise you get an instance of the first axiom schemata.

C. What you're saying is that all you need to do to determine the correctness of an equality is to see whether there are as many occurrences of 's' to the left as to the right. So, in a sense, *numbers* don't come into it.

M. Yes. And now it strikes me that the very same holds for addition. If we just consider the equalities of the form

$$J + I = K$$

for successor numerals J, I and K, then again if there are as many occurrences of 's' on the left and on the right, then you accept the equality, and otherwise you put the slash through the equals sign.

C. But why stop there? Let J be any term formed just from the plus sign and numerals, and let I be the same. Then, if there are as many occurrences of 's' in J and I, we write

$$J = I$$

and otherwise

$$J \neq I$$

and so we get the effect of *all* of the first four axioms in this very simple way which involves no *logic* at all, and no *number recognition* at all—just 'numerical comparison'—that is, seeing whether there are *just as many* occurrences of 's' on either side of the identity symbol.

M. So it seems.

C. But can this method be extended just as well to multiplication?

M. Why not? Take as an example the equality

$$s\,(s\,(0)) \cdot s\,(s\,(s\,(0))) = s\,(s\,(s\,(s\,(s\,(s\,(0))))))$$

Make a correlation the three occurrences of 's' in 's (s (s (0)))' by drawing lower lines to the first three occurrences of 's' in 's (s (s (s (s (s (0))))))'. And then do it drawing upper lines to the second three occurrences of 's' in 's (s (s (s (s (s (0))))))'. Then draw a line from the second 's' in 's (s (0))' to the lower group of lines, and then from the first 's' in 's (s (0))' to the upper group of lines. This shows by the indicated correlations that there are twice as many occurrences of 's' in the numeral to the right as there are occurrences of 's' in the second numeral to the left.

C. What you have shown me is that the equations of arithmetic could be obtained without resorting to the Peano axioms. We could instead lay down the following axioms:

S: All instances of the following schemata that result from replacing letters by numerals are axioms:

$$0 \neq s\,(n)$$

$$s\,(n) = s\,(m) \supset n = m$$

$$n + 0 = n$$

$$n + s\,(m) = s\,(n + m)$$

$$n \cdot 0 = 0$$

$$n \cdot s\,(m) = (n + m) + n$$

That is a better way of putting it.

M. In effect, the schemata yield precisely the *non-quantificational* equations, atomic and molecular, derivable as theorems from the first six Peano axioms. The axioms thus fixed are axioms for the part of arithmetic that corresponds to the arithmetic we learn in grammar school as we gain skill in counting, adding and multiplying. Let us call this non-quantificational arithmetic.

C. But **S** will not provide for such assertions as 'The sum of two even numbers is always an even number'; it will not even provide for a definition of 'even number'. So **S** is really limited.

M. That's right.

Axiomatized non-quantificational arithmetic works as follows. First, the schemata yield certain 'basic' equations by mere substitution in the schemata. All of these equations are, we might say, axiomatic. The remaining equations are obtained by identity substitutions and tautological inferences.

C. If we think of *theory* as connected with *generality*, then we could say that since generality arises only with quantification, the usual axioms—unlike the axioms by schemata **S**—provide for a *theory* of the natural numbers. The axioms by schemata **S**, by contrast, provide only for the *arithmetic* of the natural numbers. This is the core arithmetic of the natural numbers. This core is arithmetic, but includes no theory. Theory comes in with the quantifications of formulas formed from the numerals, the signs for sum and product, the equality sign, and the variables.

M. I think that is a helpful distinction.

It corresponds to a distinction between calculation and deduction. I come to know an equation or an inequality by means of a calculation. Adding and multiplying is calculating. But calculation is really insufficient for something like 'The sum of two even numbers is always an even number'. To know 'The sum of two even numbers is always an even number', I have to be able to prove it.

But let us continue to use 'arithmetic' to refer to Peano arithmetic, with the usual axioms, since that is pretty well established.

C. OK.

But now a point of metaphysical interest occurs to me. At the base of arithmetic there are only calculations. Calculation needs no ontology. So the same holds for arithmetic.

M. And here is a response: You say that at the base we find calculation. Agreed. At the base are those manipulations of signs we call adding and multiplying. Addition and multiplication, together with any other modes of calculation (e.g., exponentiation) no more require ontology than playing chess requires ontology.

However, playing a game does not give rise to statements. Addition does. For we conclude our operations with signs by saying, e.g., that 73 plus 21 equals 94. Our calculations result in statements.

This shows up in three ways. First, we use the identity sign between numerical terms. Second, we form truth-functional compounds of equations. Third, we generalize.

C. In one sense, the result of adding 73 to 21 is a *numeral*. That is, the adding consists in writing things down and what you end up with is a numeral. And so far as applications are concerned, the numeral is enough. We count the sheep and get to '73' and count the cows and get to '21'. Noting that there are no livestock other than these sheep and cows we form the statement 'There are 73 plus 21 livestock here' and then add, thereby getting to '94'. We then assert 'There are 94 livestock here'. Our new statement results from replacing '73 plus 21' by '94'.

Adding leads to a numeral it is our practice to substitute for the complex numeral from which our addition proceeds. No more is essential to the applications of addition.

M. But note that there is another sense in which the result of our addition is a statement, not a numeral. We give the result not by saying a numeral, but by saying an equation. We say such things as '73 plus 21 equals 94'. And this goes with saying that we add the *numbers*.

Applications can do without equations, but equations are indispensable to mathematics. For the applications of our calculations, the numeral is enough. For mathematics we need the equation, and equations are statements of identity.

From the point of view of applications we could register the result of our addition by saying that '94' can replace '73 plus 21'. But in relation to *mathematics* the equations are statements formulated through the *use*, not the *mention* of numerals.

C. This seems to me to be a very strong line of argument. It has great appeal. I think it touches on—is in contact with—the real sources of the idea that mathematics deals with objects.

M. Before we conclude our discussion, I was hoping we could talk about something that was bothering me.

C. What's that?

M. I am wondering if expressions such as

$$s\,(s\,(s\,(0)))$$

really are numerals. I know they are called numerals, of course.

C. What do you mean?

M. Well, look, if you ask me the number of congressional districts in Nebraska, and I tell you there as many as there occurrences of '*s*' in the expression last displayed, will I have answered your question?

C. Yes, in a way. But I would wonder why you did not just say "three". Why give me such a contorted answer?

M. There is a point I am trying to make.

That point is that if I am informed that there are as many congressional districts in Nebraska as there are occurrences of 's ' in the expression last displayed, then I know how many congressional districts there are in Nebraska—namely, as many as there are occurrences of 's ' in the expression last displayed.

If I am informed that there are three congressional districts in Nebraska, then I know how many congressional districts there are in Nebraska—namely, three.

Anyone will feel these are different cases of knowing how many.

I can know there are as many occurrences of 's ' in the expression last displayed as there are congressional districts in Nebraska without knowing that there are three congressional districts in Nebraska, and I can know there are three congressional districts in Nebraska without knowing there are as many occurrences of 's ' in the expression last displayed as there are congressional districts in Nebraska.

C. That seems reasonable. It is even more obvious when one deals with larger number of things.

The contrast seems to be one between knowing how many congressional districts there are in Nebraska and knowing the number of congressional districts in Nebraska.

M. A numeral gives one the number of things in a statement of number, for example, 'There are three apples in a box'. The terms of successor arithmetic, such as 's (s (s (0)))', serve much the same function but not exactly the same function.

C. And that is why you say the successor expressions are not numerals.

M. Yes.

CHAPTER 3
Why Axiomatize Arithmetic?

In this dialogue two characters, M and N, focus on these issues: Are the Peano axioms for arithmetic epistemologically irrelevant? What is the source of our knowledge of these axioms? What is the epistemological relationship between arithmetical laws and the particularities of numbers?

The Dialogue

N. The axioms for arithmetic, and deduction from them, are epistemologically irrelevant.

   For it is necessary, if we are to be justified in accepting this or that as an axiom, that its logical consequences not conflict with what is already known about the numbers, but instead include things already known about the numbers.

   So we cannot be justified in accepting the axioms unless we *already* know a lot about the numbers *independently* of the axioms or any deductions from them.

   After all, it's not as if mathematical knowledge *began* with Peano!

M. Your idea seems to be this: Lots of things get to be known about the numbers. Then someone proposes certain sentences as axioms. So the question arises, ought we to accept these sentences? And we decide whether we should by seeing (i) whether they logically entail antecedently known arithmetical truths, and (ii) whether they logically entail anything inconsistent with antecedently known arithmetical truths. Only if they entail what we already know, and entail nothing which conflicts with what we already know, are they acceptable axioms.

   Isn't this your thought?

N. Yes.

M. But the statements we accept *as axioms* were among the ones *already* accepted *as true*.

Peano's proposal was not that certain sentences (i) *are true* and (ii) *may serve as axioms*. Rather, he selected certain already known arithmetical truths and proposed only that they may serve as axioms.

Peano was not the first person to suppose that zero is the first of the natural numbers or that different natural numbers have different successors! His innovation lay not in *asserting* the things his axioms assert—as mathematical assertions his axioms were already familiar and accepted—but in proposing that these statements were *logically sufficient* for the whole of the arithmetic of the natural numbers.

N. Fair enough. But my point still holds good: we can be confident of the *truth* of the axioms only if their consequences, as so far known, agree with what we *already* know about the numbers, both in the sense of conflicting with nothing already known and actually leading to things already known.

For suppose someone were to propose as an axiom some sentence which logically entails something which goes against some known mathematical truth. Wouldn't we then reject the proposed axiom?

M. I find the question virtually meaningless.

Suppose that some particular equality—for example, that $5 + 7 = 12$—logically entailed something which went against some piece of mathematical knowledge already in our possession. How would you answer the question about what we would then reject?

Wouldn't you rather become suspicious of the question, or the supposition upon which it rested?

N. I don't follow you.

M. Well, we are supposed to suppose that we discover that the statement that $5 + 7 = 12$ logically entails something which conflicts with something we already know. So suppose, if you can, that it logically entails that $4 = 5$.

Now, what gives way? Do we reject the equality $5 + 7 = 12$ or the inequality $4 \neq 5$?

Suppose we reject the inequality. Then we agree that 4 = 5. But 4 + 0 = 4 and 4 + 1 = 5. Thus, since 4 = 5, 1 = 0. But then, since 2 comes right after 1, 2 comes right after 0. But 1 comes right after 0. So, 1 = 2. So, 0 = 1 = 2. And thus also 0 = 1 = 2 = 3 = 4 = 5 = 6 … That is, then all the numbers are one and the same!

Well—that makes it look as if we'd better reject the equality. Then we must hold that 5 + 7 ≠ 12. What then is 5 + 7? Well, it must be greater than or equal to 7. Suppose it is 7. Then, 5 + 7 = 7, in which case 5 + 7 = 0 + 7. But then 5 = 0. But since 6 comes right after 5, 6 comes right after 1. So 6 = 2. By the same reasoning 7 = 3 and 8 = 4 and 10 = 5 and 11 = 1, and so on. That is, *all* the numbers are just the numbers 0 through 4. Now suppose that 5 + 7 = 8. Then 5 + 7 = 1 + 7, in which case 5 = 1. But then 4 = 0, in which case there are just the numbers 0 through 3. And so on.

So the situation is disastrous on *both* alternatives.

N. I see that.

M. The same sort of thing arises in respect to the 'supposition' that some axiom entails the equality that 4 = 5, for example the axiom that zero is the first number. What would we reject if it turned out that if zero is the first number then 4 = 5? If we reject the axiom, then we agree that *some* number comes before zero. But then zero comes immediately *after* some number. So let the number be one. Then zero comes right after one. But so does two. But no *two* numbers come right after any number. So, we conclude that 0 = 2. And if that is so, then 0 = 3 = 4 = 5 = 6 = 7 = 8 = 9… just as well. And if it isn't *one* that zero comes after, it is *some* number—and the conclusion will be that the numbers then stop with *its* immediate successor, namely zero!

So, if we reject the axiom we will have to accept something outlandish, such as 0 = 2. If we do not reject the axiom we will have to accept something equally outlandish, for example that 4 = 5.

So, and please do not take this remark badly, it seems to me that no *thinking* has gone into this idea that we must test the axioms against what is already known. It has just been words without thought.

N. Perhaps there is something in what you say. I do now feel as if I never bothered to think through the supposed supposition I was making. But what if someone proposed something *brand* new as an axiom—then, surely, it would have to meet the test of agreeing with what was already known in arithmetic, and so couldn't be a *source* of knowledge.

M. But what is this 'something brand new'? Presumably it will be a sentence of arithmetic. So it won't be *syntactically* or *semantically* novel. So it must be one whose *truth–value* is as yet unknown to us.

But *no one* proposes as an axiom for arithmetic some arithmetical sentence we aren't already sure of!

N. But surely this *could* be done. Didn't Gödel suggest that we might try to think up further axioms for set theory? And any such axiom would not really be *known* to be true at the start? (Gödel 1964, pp. 263-265)

M. Perhaps. But right now we are talking about axiomatic arithmetic, not axiomatic set theory. The two may be quite different, and certainly the intellectual pressures which gave rise to them are entirely distinct. So let's just stick to arithmetic. And here my point is this: In arithmetic no one proposes as axioms statements about which we are as yet *uncertain*.

N. As I reflect on what you say, I find myself forced to agree. The actually proposed axioms, and very likely any that anyone would seriously propose, are statements about whose truth we are already satisfied. In a word, and this is the point I did not appreciate, axioms are selected from what is *already known*.

I think that what happened in my thinking was that I slipped from asking what would be needed to accept a statement *as an axiom* to asking what would be needed to accept the statement. Then, since a condition on its being an axiom is that a statement, together with the other axioms, leads to all the already recognized truths, I slipped into thinking that this was a condition on accepting the statement itself—as if it needed to pass a test taken simply as a statement of arithmetic!

In any case, I now seem to see that it is quite correct to say that the statements which are proposed as axioms for arithmetic need pass no special test since they are all among the statements already known to say true things about the numbers.

But though I will grant that proposed axioms state about the numbers things already known, surely there is much in addition to what is said by those axioms that also is *already* known about the numbers. And those further, *already known* things are not known by *deduction* from the axioms. That point remains secure.

And so I will continue to assert that the axioms and deductions from them are epistemically irrelevant.

M. But if other things in arithmetic *could* be known by deduction from the axioms, then if we can see how we *do* know that the numbers are as the axioms state them to be, then we can see how we *can* know all the rest. We could gain a certain insight into the *knowability* of the propositions of arithmetic, even if not into the manner in which all or some of them have actually become known.

N. But it well may be that we couldn't know that the numbers are as the axioms state them to be *without* already knowing ever so many of the more specific points about particular numbers which follow from the axioms. And in that case we will not gain the insight to which you refer.

I will make my point this way: You assume that the axioms can be known *on their own* and apart from knowing anything else pertaining to the numbers. But this assumption may be in error.

M. Is that my assumption? It may be a kind of psychological impossibility to know that zero comes before all the other numbers and *not* know that it comes before one. For it might be impossible for us *not* to see that zero comes before one if it comes before all numbers other than itself. If so, we could not recognize that zero comes first without recognizing as well that zero comes before one. So it might be impossible to know the truth of the general proposition without knowing the truth of some of its instances. I have not suggested or assumed the opposite.

But that would not show that knowledge of the truth of the general proposition is *based* on knowledge of the truth of its instances.

N. On what else might it be based? Can't we see that it *must* be based on knowledge of the truth of at least some of its instances?

M. I am not at all sure on what our knowledge of the truth of the axioms is based. But I think I can see that it is *not* based on such knowledge of the truth of its instances as we may possess. And here is how I see the matter. Consider the axiom that states that zero precedes every other number. Its instances are 'zero precedes one', 'zero precedes two' and all the rest. Now, we know the truth of only finitely many of these instances. But we surely couldn't know that zero precedes *every* other number by knowing that it precedes three or fifty or fifty-thousand other numbers. So our knowledge of the truth of the axiom cannot be based on our knowledge of the truth of its instances.

If we came to believe that zero precedes every other number by noting that it precedes one, and two, and three, ... , and five million, then that would be merely a *conjecture*—not a piece of knowledge. And a very weak conjecture at that, since we would be extrapolating from the finitely few to the infinitely many. This is actually a point C made in our last conversation.

Yet, we surely *do* know that zero precedes every number other than itself. And so it seems that this knowledge *is* independent of such knowledge of the truth of its instances as we may possess. And the same holds for each of the other axioms. So they all are known *independently* of their instances. But their instances can be deduced, and thus known through them.

So that is *one* way in which knowledge is possible in arithmetic, and *this* way of knowing will become fully clear to us as soon as we make it clear to ourselves how we *do* know the truth of the axioms.

N. I feel very uncomfortable about this. It continues to seem clear to me that we know, e.g., that zero does not come after one, and know that every bit as much as we know that zero follows no number and do *not* know it by deducing it from that general proposition.

M. But I have said only that by reference to the axioms we can arrive at an understanding of how arithmetical knowledge *can* be obtained—not that we will arrive at an understanding of how arithmetical knowledge *must* be obtained, or even *has* been obtained.

It may be that we *can* know that zero comes before thirteen by deducing that proposition from the known truth that zero comes before every other number, and that is consistent with our *actually* knowing that zero comes before thirteen *independently* of any such deduction.

I am not suggesting that we ordinarily *do* know, much less than we *must* know, the familiar sums and products *by deduction* from *independently known* axioms. I say only that we *can* know these sums and products by deduction from independently known axioms.

N. I see what you're saying, and I don't see how to object to it. Yet there seems to me to be *something* wrong in your conception of the matter.

It seems so clear to me that knowledge of the truth of various instances *comes first* and that knowledge of the truth of the generalization *comes second* and *couldn't* come first. And so I feel that knowledge of the truth of the axiom really is secondary or derived. Despite your arguments, I cannot shake myself of this feeling.

M. Well—you *say* this, but then you don't explain it. So I am quite at sea about how best to respond to you. I really don't see why you say we *couldn't* know the truth of some consequence of an axiom by deducing it. Deduction surely is *one* way of obtaining further knowledge from some already obtained knowledge. I hope you will not go so far as to question *that*.

N. That has tempted me. But let me try this instead.

As I understand it, people had a good grip on a lot of arithmetic *prior to* getting a handle on zero. This shows that people could know—*did* know—enormously much about the numbers *without* so much as *conceiving* the content of most of the axioms—for most of them include the idea of zero.

M. Agreed. But that would be a somewhat different system of arithmetic, and so would have somewhat different axioms. Instead of asserting that zero comes first among the numbers, the right axiom for this arithmetic would assert that one comes first among the numbers.

But what turns on your point? I have not denied that our actual knowledge may be independent of deduction. I have only asserted that deduction from antecedently known axioms is *a* way of knowing.

N. But the important point is that it is possible to have *fragmentary* arithmetical knowledge. And just as there is arithmetic without zero, there can be arithmetic without generalization. People can know a lot about individual numbers—their sums and products— without as yet so much as *considering* generalizations like 'Zero comes after *no* number'. It might not be the case that anyone who *considers* that statement fails to recognize its truth. But people might be, as it were, oblivious to it and *still know lots of things about lots of numbers.*

So I will now put my point this way: Our knowledge of the individual numbers—as contrasted with our knowledge of arithmetical laws—in *no* way depends upon our knowledge of those laws.

M. But now you are no longer viewing the epistemic irrelevance of axiomatic arithmetic in the same way that you did at first.

N. I agree. My initial 'view' made no sense at all. It was, as you put it, words without thought.

But that recognition did not free me of a feeling that somehow the laws of arithmetic are *secondary*. And so, I am now working on *that* idea.

What I now wish to say is that the fact that we can know enormously much about the numbers without so much as considering the laws shows the irrelevance of those laws to our basic arithmetical knowledge.

M. In a way what you say seems undeniably correct. But aren't you still missing my main point—for I have only said that it is *possible* to arrive at arithmetical knowledge by deduction from axioms independently known to be true. Not that we must or even actually do so arrive at arithmetical knowledge.

N. Yes. For some reason I keep slipping up on that point. Still—there is something I am after, so let me shift ground again.

   What I now want to say is that knowledge of an axiom *does* depend on knowledge of its instances. Not all of them, but some of them. We first come to know instances. We perceive a pattern in the instances. We then generalize on the known instances to bring out that pattern. I do not mind calling that a kind of speculation. Then by deduction we bring out further particular points that we can then put to the test.

   This, I now think, is how things work in arithmetic. We know the axioms to be true only by seeing that again and again their entailments are true.

M. Here you're viewing the axioms as analogous to expressions of empirical regularity. This much is right in that analogy: the axioms are laws. They are laws of numbers.

   You're also suggesting that the laws of number are hypotheses suggested by particular facts and then tested by seeing what further particular facts they lead to.

   This makes knowledge of the laws entirely secondary.

   But I feel that *in some way* knowledge of laws is present in even the most elementary pieces of arithmetical knowledge—in a way in which knowledge of an empirical generality is not involved in knowledge of its instances.

   I could put how I feel about the matter this way: I need not in any way work with the idea that all ravens are black to spot the blackness of *this* raven.

   So I just cannot accept the sharp distinction you draw between knowledge of the laws of number and knowledge of the numbers.

N. That is extremely obscure.

M. I agree. And I could say even more obscure things than that on the topic! But don't worry about the obscurity, for even an obscure remark might lead to better things.

So let me go on and try this assertion, that there is something *general* involved in what goes into knowing anything arithmetical.

N. I would say that you were true to your word, you can indeed increase the obscurity of your pronouncements!

But seriously, tell me what you have in mind—if you *do* have anything in mind.

M. Well, something general comes into learning the numbers— learning which numbers are which and what the sums and products are. For you don't just learn this number and then that one, in a random fashion, as if you were learning about this raven and then about that one, but you learn a *system*. You learn to go from one to two, and from two to three, *and so on*. You need to do something like master a rule. And there is something general about a rule.

I would not deny that this is still very obscure. But it feels right nonetheless, and perhaps just a bit less obscure than what I began with.

N. But don't some people count one, two, many? Or a child might master just the numbers through five.

M. Yes. There are these fragments of arithmetic which, taken on their own, are in some ways like and in more ways unlike arithmetic. But if you've got the *numbers* then you don't have just five or fifty of them. In a certain sense you have them *all*. And that's where the generality or law comes in. We might say that you master a law of construction for the numbers, or for recognizing them. And it is the mastery of *that* law which gets reflected in the first two axioms.

A person might be thrown by the question 'And is there a greatest number?' while yet knowing full well that if there is one more swan than fifty there are fifty-one swans, and if there is one more swan that fifty thousand, then there are fifty thousand and one swans. *And so forth.*

And the person thrown by the statement that one comes before every number other than itself will nonetheless always start with one when counting! Here, we might say, a certain feature of his or her practice constitutes a kind of recognition that one comes before all the other numbers—and it is this feature of *practice* which is explicitly set out in the axiom.

Even a child will not be said to as yet know its numbers if its counting-like actions begin at times with one, and at times with three, and so on. Counting isn't just any kind of repetition of familiar sounds together with gestures of pointing.

N. But surely miscounting is counting. Just as invalidly inferring is inferring.

M. A child can miscount only if it has mastered the technique of counting. The very acts that with us are miscountings would not be that if they constituted the normal or typical performance of a child.

But we are drifting. Let's review where we're at.

We've moved from talking about axioms and their instances to the topic of the general and the particular in arithmetic. The question we are now considering could be put as follows: Is knowledge of the laws of number somehow *implicit* in our knowledge of the particular numbers and their sums and products?

I am inclined to think this is so. You are inclined to doubt it. You feel that knowledge of the numbers—which are which, and what the various sums and products are—comes both first and quite independently of knowledge of laws. I feel that knowledge of laws is implicit is our knowledge of the numbers.

I feel that laws are implicit even in an arithmetic entirely lacking the means for putting laws into words. This is connected with the importance I attach to the *systematicity* of arithmetic. But you feel that knowledge of which numbers are which, and what the sums and products are, are points of information which can be picked up one by one, in a non-systematic way, and that a system emerges only as we see certain patterns in the particular number facts we arrive at one by one.

Do I get our two perspectives right when I describe them in this way? What do you think?

N. Yes. I think so. And now it strikes me that you have perhaps moved away from your earlier view that arithmetical knowledge *could* be based on knowledge of axioms and deduction. For now you see the axioms as there in arithmetic even when the arithmetic in question cannot so much as formulate them as axioms. For what you say of laws holds for axioms, since axioms are laws.

M. Yes. It no longer is clear to me that mathematical knowledge comes down to knowledge of laws plus *deduction* from laws. The knowledge of the laws is, as it were, there from the start as knowing *how* to count, to add, and to multiply.

I would now prefer to say that talk in terms of axioms and deduction no more than *alludes* to what is going on in obtaining arithmetical knowledge—or *represents* that knowledge in a certain way. It puts things, we might say, in a 'deductive style'. And that might even be highly misleading presentation of that knowledge.

N. Well—I'm certainly pleased that you've finally become quite clear in your statements!

M. Yes. I too thought I was doing a lot better.

So here we are. Confused. Not a bad starting point for philosophy.

There are the *particularities* of the numbers, and the *laws*. The real issue, as I now see things, has nothing in particular to do with *axioms*. They are of interest only when one wants, as it were, to logically sum up a body of knowledge. Instead, it is the relation of the laws to the particularities which concerns us, and how law and particularity connect up with knowledge.

Or I might put it this way: We are wondering about the relation between non-quantificational and quantificational arithmetic.

So we need to ask ourselves: What are the roles of generality in arithmetic?

And here three things immediately present themselves: First, there is the generality present in our methods of calculating particularities. Second, there is the generality that enters into the formation of new mathematical concepts, and, third, there is the generality we arrive at through proof.

CHAPTER 4
A Conversation about Numbers and Knowledge

P, a Platonist, takes up the conversation with M. In this conversation reference is made to the first two axioms for arithmetic. The first two seem most important for the so-called infinity of the natural numbers. The first axiom states that zero is the immediate successor of no number (in symbols '$\forall x \, (0 \neq s \, (x))$'). The second states that different numbers have different immediate successors (in symbols '$\forall x \, \forall y \, (x \neq y \supset s \, (x) \neq s \, (y))$'). In the course of the conversation (as in all the dialogues in this book) the term 'number' is taken to cover just zero and the successors of zero. So in this context 'number' means natural number. It probably is worth mentioning that the second of these axioms will not, on its own, provide for the infinity of the natural numbers. To see this, interpret '$s \, (x)$' to mean 'the person x is'. Then the axiom says that if x isn't y then the person x is isn't the person y is. And this is true even though there are just finitely many persons. A similar point holds for the first axiom. For if we interpret 's' to mean 'the mother of' and '0' to stand for Quine and the variables to range over people, the axiom says that Quine is not his own mother. And this too is true though there are only finitely many people. So, neither axiom on its own logically suffices for the infinitude of the natural numbers.

The Dialogue

M. How many numbers are there? I mean the so-called natural numbers—you know, zero, one, two, three and the like. How many of them are there?

P. Infinitely many.

M. I find it striking that you feel you both understood the question and know its answer. But seeing that you have answered as you did, can you also tell me how do you know that there are infinitely many?

P. Everyone recognizes that there are infinitely many natural numbers.

M. Perhaps. But if I asked you how you knew some dogs bark, your answer would not be that everyone knows this: You'd just say how *you* knew it, or you'd tell me how it can be known. And if I told you I wanted to find out for myself whether or not any dogs bark, you'd advise me to go to a shelter and visit the dogs (and adopt one while I'm at it). I'd then be able to hear for myself whether any of them bark.

　　So let me ask you again. How do you know there are that many numbers? Or just tell me how I might find out about this for myself.

P. Well—the usual axioms say there are infinitely many numbers. More specifically, the first two are true only if there are that many numbers.

M. So, if I can find out that those axioms are true, then if I can see as well that they are true only if there are infinitely many numbers, then I shall be in a position to know that there are infinitely many numbers. This sounds reasonable, and it starts with the truth of the axioms.

　　So tell me, are the axioms true?

P. Of course.

M. And is this something you know?

P. I think so.

M. Then tell me, what has shown you that they are true? How did you come to know this?

P. Well, you're asking me how I know that zero is not the immediate successor of any number and that different numbers have different immediate successors. For that is what those two axioms state. I certainly think that I do know these things. And so I know that zero is not its immediate successor or the successor of its immediate successor. But these numbers can't be the same, since if they were they'd be successors of the same number, in which case zero would be its immediate successor. But it isn't, since it isn't any number's immediate successor. So there are at least these three numbers: zero, its immediate successor, and the immediate successor of its immediate successor. Reasoning in this way can show that there are infinitely many numbers, that is, successors of zero.

M. But look: I asked how you came to know that the axioms are true, or what shows they are true. But you never said anything about this. You just went on to *use* the axioms in arguing for other points. Isn't that so?

P. Yes. I got ahead of myself.

M. And I also think there's a mistake in what you actually said, right near the end—when you said that by reasoning from the axioms you can show there are infinitely many numbers.

P. How so?

M. Well, suppose you actually *knew* that zero is not its immediate successor and that different numbers have different immediate successors, as the axioms state, then you'd be in a position to know that zero isn't its immediate successor and that its immediate successor is neither it nor zero, and the like. But you don't by such deductions come to know there are infinitely many successors of zero. I mean, you come to know that there is one, and that there are two, and that there are three, and so forth. But you never get to know there are *infinitely* many. Isn't that right?

P. Well, I grant that you don't literally get to there being infinitely many in that way, but can't you show from the axioms that every number has an immediate successor? And doesn't that amount to the conclusion that there are infinitely many of them—if each has a successor?

M. But the proof that every number has an immediate successor is trivial and doesn't even involve the specifically mathematical axioms. You surely know how it goes. You first write down '$s(n) = s(n)$' by identity, and then by existential generalization you write 'There is an m such that $s(n) = m$', which says that every number has an immediate successor. So, for *that* you don't need any axioms beyond the logical ones!

In fact, the statement that every number has an immediate successor is even *logically compatible* with there being just *one* number. So, that statement seems *not* to say that there are infinitely many numbers!

P.   OK. I think I see that—though in some strange way I had somehow thought it did say that there are infinitely many numbers. I know I had always taken it that when you prove that every prime number is succeeded by some prime number, you prove the infinity of the primes. But now I see that even this statement about primes doesn't *logically* entail that there are infinitely many primes. But now, as I reflect on what I've just said, it isn't clear to me that I should conclude that the statement doesn't say there are infinitely many primes. Perhaps it says, or implies, more than it logically entails?

After all, the statement that Harry is older than Bill implies that Bill is not as old as Harry, but certainly doesn't logically entail that Bill is not as old as Harry.

M.   But what inclines you to say that the statement that every prime is succeeded by some prime implies the infinity of the primes? Isn't it that you're already thinking of the numbers as an unending series of successors of zero?

After all, if you as yet had no idea whether or not zero and its successors were one and the same, then, for all you would know, it could be a perfectly vacuous truth that every prime would be succeeded by some prime. For that would be true were there no primes at all.

Or if you thought that the successors of zero were distinct through eleven and that all successors of eleven *were* eleven, then since eleven is prime, every prime would be succeeded by some prime, namely *itself*.

Doesn't this show that it is only because you *already* think of the successors of zero in a certain 'unending' way that you take the statement about primes as indicating the infinity of the primes?

P. Yes. I see that. But isn't it part of the very *meaning* of 'immediate successor' that the immediate successors of zero form an infinite series?

M. Well—suppose it is. Then the question becomes this: Does zero, or any other number, have an immediate successor? And it is hard to understand what it would mean to say that it is part of the very meaning of 'immediate successor' that zero has one. At best you could say that it is part of the meaning of 'immediate successor' that *if* zero has one, it is not *zero*.

One might even make the following definition, that n immediately succeeds m just in case n succeeds m and $n \neq m$ and there is no k such that n succeeds k and k succeeds m. Given that, the statement that zero has an immediate successor will *logically* entail that this immediate successor is not zero. But it is no part of the meaning of 'immediate successor', as just defined, that any number has an immediate successor.

I just don't have any clear idea of how one might, as it were, 'get' the infinity of the numbers out of meaning.

P. I think I am again persuaded by what you say, though I feel a great deal more might yet be said on the topic. But it does seem to me right that the statement that every number is succeeded by some number will seem to assert the infinity of the numbers only if we're *already* thinking of the series of successors as unending.

So now the problem is to actually put that thought, the thought that there are infinitely many natural numbers, into the words of arithmetic.

M. But why set ourselves this problem? Don't we already have a perfectly acceptable way of saying what we want to say, I mean by the very words we've been using—for example, that there are infinitely many natural numbers.

P. My feeling is that the word 'infinite' is hardly more than a word for enormously many or unimaginably big in its *non*-mathematical uses. It is in mathematics, and more specifically in connection with the mathematics of the natural numbers, that infinity first shows itself for what it is. And so it seems to me that until we determine just how the infinity of the natural numbers expresses itself in the mathematics of the natural numbers—or perhaps more generally in set theory—we shall not have a clear view of the topic.

And not just that. Since we have the axioms for the natural numbers, and can prove many truths about the natural numbers from these axioms, if we can express the infinity of the natural numbers in the language of those axioms we stand some chance of also *proving* that infinity *from* the axioms. But if the terms in which we express the infinity of the natural numbers go *beyond* the terms used in the axioms or what is definable by reference to those terms then the axioms certainly will be useless for proving the infinity of the natural numbers, and hence cannot help us to *know* that there are infinitely many of them.

So, because of these two considerations I think it best or even essential, that we seek to determine how infinity is expressed within mathematics—and that means, in the language of axioms.

M. That seems to be a good answer. And I suppose that if it were to prove impossible to express the infinity of the natural numbers in the language of the axioms, that would be important to know.

And then we would want to inquire about how we might *extend* the language of the axioms to arrive at what we need. And this too, should it prove to be the case, would be important to know.

Perhaps the only point in what you say which seems really questionable is your assumption that we might come to know something by proving it from the axioms. But that is a very difficult and obscure topic—I mean the relation between axioms, deduction and knowledge—and one we had best not pursue if we are to continue the main thread of our conversation.

So let's go ahead as you suggest, seeking to express—or grasp—the infinity of the natural numbers in terms of the language of the axioms.

P. Good. But now, if the sentence saying there is no greatest number doesn't say that there are infinitely many numbers, what does? And surely *something* does. For it is easy to see that given the usual axioms we can prove any sentence of the form '$n \neq m$' so long as *n* and *m* are different numerals—either '0' alone or '0' preceded by one or more occurrences of 's'. And that means we can prove *infinitely* many of them.

M. You mean that each of infinitely many is provable. Not that you or anyone else might carry out infinitely many proofs. Is that right?

P. Of course. But all those non-identities are *provable*. And there are infinitely many of them. And so the axioms do show that there are infinitely many numbers. The question is this, which provable arithmetical sentence *expresses* that fact.

M. Perhaps the answer is that *no* strictly arithmetical sentence expresses the infinity of the natural numbers. After all, to express this, according to the experts, you would need a sentence that says that the *set* of successors of zero is one-one with one of its proper subsets. But no such sentence is available in the language of the axioms, for no signs specific to set theory occur in those axioms.

P. Now you seem to be implying that the axioms do not insure that there are infinitely many successors of zero because the language in which they are formulated lacks the means for asserting their infinity.

But, as I said at the start, they do insure that. For the axioms won't be true unless there are infinitely many successors of zero. So I am not sure what to make of what you say.

M. I think it might help us if we were to distinguish between what can be proven *from* the axioms, and what can be show by *reflection on* the axioms.

P. Could you explain what you have in mind?

M. I hope so. To begin with, when we speak of *proving* something from the axioms, what we have in mind is logical deduction, and what gets logically deduced is a logical consequence. So let's take 'proof' to refer to that. Then I think it is natural to distinguish between those logical consequences of an axiom that do not in turn logically imply that axiom, and those that do. The logically entailed but not logically entailing sentences are the ones which, as it were, spell out the content of the axiom. The remaining logical consequences, by contrast, could be said only to reformulate that content.

P. That seems right.

M. But then notice that from the two axioms we are talking about, the ones which deal specifically with the sequence of successors of zero, and considering only the sentences which spell out the content of those two axioms, we can prove—that is deduce

$$0 \neq s\,(0),\ 0 \neq s\,(s\,(0)),\ s\,(0) \neq s\,(s\,(0)),\ \ldots$$

thereby expressing that those successors are all different one from the other, and we can prove as well various conjunctions with these sentences as their conjuncts. But that is all. And each of the conjunctions, no matter how many conjuncts it may have, asserts the distinctness of only *finitely* many numbers, and each of their existential generalizations will assert the existence of only *finitely* many distinct numbers. So, by proving from the axioms the things which spell out the content of the axioms you never get to any sentence saying that there are infinitely many numbers, infinitely many successors of zero. Isn't that so?

P. Certainly. I believe this is just the point you made earlier.

M. Indeed.

P. And so we should say that reasoning from these axioms so as to spell out their content can't get us to the infinity of the natural numbers.

M. Yes, that is just what I had in mind. But when we *think about* or *reflect on* the axioms, as we are now doing, we can see that we can, by deductive reasoning, spell out their content in terms of infinitely many such sentences. Not that we can actually execute all those deductions, but that we can see how they *could* be carried out. Isn't it by seeing just this, that infinitely many sentences spelling out the content of the axioms are deducible from those axioms—the very thing which we take ourselves to have just seen—that we are convinced that the axioms hold only if there are infinitely many successors of zero?

P. Yes.

M. Well then, we certainly can see something about infinity from *thinking about* the axioms, and thus, in a sense, 'from' the axioms. But not by *using* the axioms to deduce this or that—and thus not so much from *the axioms* as from *reflection on the axioms*.

The axioms themselves don't show us that there are infinitely many successors of zero. And if we come to see that the axioms are true only if there are infinitely many successors of zero, then what shows us that is a certain type of investigation of the axioms.

It is all the difference between what you get by deducing things and what you get by observing that they are deducible.

P. Is that really so? Let's conjoin the first two axioms, the ones which say that zero is the immediate successor of no number and that different numbers have different immediate successors. They are the only ones that are relevant to our discussion. Then we have a single axiom that is true only if there are infinitely many successors of zero. And now you seem to be saying that this conjunctive axiom doesn't assert that there are infinitely many successors of zero since it asserts only as much as can be spelled out by certain deductions from it.

But how can something be a condition of the truth of some sentence and *not* be asserted by that sentence?

M. Let me begin to answer your question by asking yet another question. Suppose I were to now ask you what you mean by the form of words 'there are infinitely many successors of zero'. What would you answer?

P. Just what we said before, that the set of successors of zero is equinumerous with one of its proper subsets.

M. But what then of the point I raised earlier? How can *that* be asserted by an *arithmetical* sentence? For an arithmetical sentence actually lacks the vocabulary for formulating that assertion!

After all, what a sentence asserts is like what a person believes, and not just *any* form of words can be used to say what someone believes. If what a person believes is that snow is white, then even if snow is white if and only if snow has such and such a surface molecular structure, the person who believes that snow is white doesn't thereby believe *that*. And the same holds for the *sentence* 'Snow is white'. Even if what it asserts is true if and only if snow has that surface structure, the sentence doesn't *assert* that it does.

Someone who believes that zero comes first in the natural numbers and that different natural numbers have different immediate successors doesn't *thereby* believe anything at all about equinumerous sets—even if those axioms are true only if the set of natural numbers is equinumerous with one of its proper subsets. And the same holds for the sentences that assert what he believes. They too don't assert anything at all about equinumerous sets.

P. That seems right, though it too is something we might talk about another time at greater length. For it has never been entirely clear to me just what it is for something to be a truth-condition for a sentence, and what relation its truth-condition has to what a sentence says.

But however this may be, I still insist that *something* having to do with the infinity of the natural numbers is asserted by those axioms. So though I am inclined to say that the axioms don't assert that the set of natural numbers is infinite, I don't feel it is right to say that they say *nothing* at all about the infinity of the natural numbers.

To be honest about it, though, I feel quite confused at this point.

M. Don't suppose that I feel perfectly clear on these matters. Far from it. I have my own doubts about what we're saying. But let's push on.

I think you agreed that none of the non-identities deducible from that axiom assert or imply that there are infinitely many successors of zero.

P. I did, and still do.

M. So the axiom doesn't provide a proof that there are infinitely many successors of zero in *that* way. How then does it provide the proof? Shall we say it does so by providing for the deduction of a sentence that is true only if there are infinitely many successors? But then the axiom *itself* is such a sentence, and so no deduction is necessary. Shall we then say that what is needed is the deduction of some sentence which says *in so many words* that there are infinitely many successors? But to say that it seems you need to talk about *sets* of successors, and the terms for saying that are lacking in the language of the axioms.

So, you must say that the axiom *doesn't* yield a proof that there are infinitely many successors of zero. This is what I've been arguing. And, further, in the customary way at least it also doesn't assert that there are infinitely many successors of zero.

As to whether or not the axiom *in some sense or other* asserts that there are infinitely many numbers, well I am not sure just what to say on that score.

But if we agreed that the axiom asserts something along the lines of the infinity of the natural numbers, then we shouldn't be surprised at the result that the axiom asserts what it cannot serve to prove. For that really is not a surprising thing to say on any account.

P. I agree. And now I can see that I made a slip when I said that you made it look as if the axiom doesn't *assert* anything pertaining to the infinity of the natural numbers. For all you said, what that the axiom doesn't itself show is that there are infinitely many natural numbers. That hardly rules out its *asserting* that point. And, as you've rightly observed, a sentence cannot serve to establish or to prove what it asserts.

But does the axiom in question—I mean the conjunction of those two axioms—assert that there are infinitely many successors of zero? You left that open. But it still somehow seems right to me to deny that the axiom asserts this—for I think it cannot assert more than is asserted by those of its logical consequences which are not equivalent to it—for it is just these consequences which spell out its content.

But even as I say this I still feel that the axiom somehow really does say that there are that many successors of zero. I really feel confused.

M. Look, we've noted that among the logical consequences of that conjunctive axiom which spell out its content *none* asserts the distinctness of more than some particular number of successors. But couldn't you say that the axiom nonetheless asserts something beyond what is asserted by those sentences? For it asserts all of them, or, rather, asserts *everything* that they assert. And, of course, none of them asserts *that*! So you cannot rightly argue that since none of the sentences spelling out the content of the axiom says there are infinitely many numbers, the same holds for the axiom.

P. Yes. That is just the right thing to say! For just consider the sentence 'All swans are white'. Each of its instances says less than it says. But *it* says *everything* they say, and that is said by none of them. The point is so obvious as soon as you notice it!

M. Yes. That is often the way it is in our thinking. We fall into confusion because we fail to notice the obvious.

But let me see if I can describe where we find ourselves for the moment. It now seems right to us to say that though the axiom certainly cannot be used to *deduce* that zero has infinitely many successors, it nonetheless does *assert* that this is so—for it entails, for *each* number *n*, a sentence which asserts, in effect, that there are *n* different successors of zero. In that way it covers all the infinitely many cases. And what the axiom itself does is to *sum up* all these cases and thus says what no *one* of them can say, that there are infinitely many successors of zero.

But to see that it asserts this—to *show* that it does—it plainly is not enough to *reason from* it. One will also have to *reason about it*. And that would be a piece of 'metamathematics' and would consist precisely in *proving* that a set provides a model for this axiom only if it is equinumerous with one of its proper subsets. So the bit of metamathematics in this case would be model theory carried out in some theory of sets.

Does that set out our current view correctly?

P. Almost. But though I agree that this bit of metamathematics is available, I would add that we might not need to avail ourselves of it. I don't mean that we can avoid *reasoning about* the axiom if we are to *see that* it asserts all it does assert, but it is not clear to me that such reasoning must be carried out in model theory or *must* appeal to sets. I think one might simply realize that the axiom has, for each number *n*, a consequence asserting the existence of *n* distinct numbers.

M. I'm not sure what it is you're driving at.

P. Well, we have said that the conjunctive axiom asserts that there are infinitely many successors of zero. But it is said that to assert this—namely, that the axiom asserts there are infinitely many successors of zero—you need to bring in sets and say that the *set* of all successors of zero is equinumerous with one of its *proper* subsets. So that makes it appear that if we are to see that the axiom asserts the infinity of the natural numbers we must 'progress' to set theory. This is what seems doubtful to me.

M The experts on these matters tell us that there are just two ways of asserting that a set is infinite—either we say it has a subset equinumerous with the set of natural numbers, or we say it is a set equinumerous with one of its proper subsets.   And isn't something along these lines correct? Or would you disagree?

P. I have no problem with that assertion. To say that a *set* is infinite may well be to say one or the other of those two things. But must it be the same to assert that there are infinitely many successors of zero and to assert that the *set* of successors of zero is infinite. The latter claim surely brings in sets—for it is couched in the very language of sets. But the former claim is about the successors— not about this or that set.

M. But I thought it was only by bringing in sets that *sense* could be made of talk of the infinite.

P. I am not sure what to make of that. But I think I know that the conjunction of those two axioms does not assert that the set of successors of zero is infinite, and yet says what can be true only if there are infinitely many such successors.

M. I don't know what to make of that. You seem to be contrasting the way we express the infinity of the natural numbers in *arithmetic*, with the way we express that infinity in *set theory*. Is that what you're up to?

P. Yes, something like that. If, as it were, you *stand within arithmetic* you arrive at the infinity of the natural numbers by thinking out the conjunctive axiom. But if you stand within set theory you arrive at the infinity of the natural numbers quite differently, by thinking out the statement that the set of natural numbers is equinumerous with one of its proper subsets.

M. Is it your idea then that different infinities are expressed in arithmetic and set theory?

P. I am not sure what to say. When I think through the axioms I seem to be thinking that each next successor of zero is distinct from the earlier successors of zero. That, at any rate, is what the axioms enable me to prove.

Given a successor, I can sketch out all the earlier successors by dropping one 's' after another until I get to zero, and then use the conjunctive axiom to prove that the given successor is different from all of those, zero included. And I see that I can always move to the next successor and then again show it to be different from the ones earlier than it.

That's how my thinking seems to go in arithmetic.

But in set theory I begin with the assertion that there is a set of a certain kind—for example, the least set N including the empty set and for each of its members, the unit set of that member—and then prove that there is a set K of disjoint pairs such that for each member x of N there is a member y of some proper subset M of N such that $<x, y>$ is a member of K, and conversely. So—1 come to see that the set N has the property which marks it as infinite—that of being equinumerous with one of its proper subsets.

Here what I arrive at are certain infinite structures, whereas in the purely arithmetic case what I arrive at is a quite different recognition.

M. I think the last remark is really very unclear. But still, I get a sense of how you're looking at things from what you say. Tell me, would this be a way of putting your idea, that in arithmetic we express, or arrive at, only a potential infinity, whereas in set theory we do not stop short of actual infinities, or completed infinities? I have heard such terms used, even among mathematicians of this century.

P. Yes. Those terms—potential and actual—may rightly hit off the difference I seem to see in the two cases.

You see, what seems clear to me is that the axioms do assert *infinity*, if I may so put it, but obviously do so in a way quite different from that in which we assert infinity in set theory. We might say that the axioms *present* the numbers as limitless, whereas the sentence from set theory states that the numbers are limitless.

I am not at all sure that this form of words hits it off right. But there must be some way of drawing the contrast that I seem to be seeing.

M. Well—there does seem to be a difference. Set theory isn't arithmetic and arithmetic isn't set theory.

But tell me, do you think the idea of infinity is somehow better expressed in set theory than in arithmetic?

P. In a way, yes. But I meant to contrast only the *axioms* with set theory—for I am not ready to agree that arithmetic isn't set theory, even though it isn't the whole of set theory; it certainly is a part of set theory.

M. How queer. Especially since set theory didn't even exist until the mid-1800s!

P. That's just a bad joke. All I meant was that arithmetic can be formulated in set theory.

M. So be it. But then look at the sentences of set theory which provide a set-theoretic formulation of the axioms for arithmetic. They also fail to include or entail sentences asserting that the set of successors of zero is infinite. For the axioms giving the set-theoretic formulation of arithmetic make no mention of a set of natural numbers, or any other infinite set, for that matter. Rather, reference is made just to finite sets built up from the empty set. Of course, the surrounding set theory has further resources and often is so formulated as to contain an axiom asserting there is a set that then can be shown to be infinite—the so-called 'axiom of infinity'. But what I want to stress here is that the contrast between the arithmetic formulation of the infinity of the natural numbers and the set-theoretic formulation of that infinity could be drawn all over again within set theory.

P. I suppose so, but so what?

M. I just wanted to indicate that the move to a set-theoretic formulation of arithmetic doesn't alter the fundamentals of the situation.

But now, we can also ask whether there really is a contrast here, and, if there is, what kind of contrast is it.

P. How do you mean?

M. Well, one alternative is that it is the same notion of infinity that gets expressed in these two ways. Then, presumably, there is no choice between these two expressions in the sense of one getting something right which the other gets wrong. So there might be no contrast in terms of right and wrong. On the other hand, it may be that *different* notions of infinity get expressed. And again there will be no contrast in the sense of one getting something right which the other gets wrong.

P. But look, might it not be that though both expressions, the arithmetical one and the set-theoretic one, present the same idea, one does it more clearly or completely than does the other? Certainly there can be a better and worse *formulation* of some idea. On the other hand, as you suggest, it might be that the ideas are quite different. But then it might be that one is right and the other is wrong. After all, can't ideas be right or wrong?

M. I don't follow you.

P. Then let me put it this way. It might be that the mathematical infinite is an actual or completed infinity. In that case only set theory *fully* captures it. The axioms for arithmetic only *partly* capture it. That is, it might be that from within arithmetic the actual infinity of the successors of zero can be seen only as the *possibility* of always getting to a provably new successor. But with the move to set-theory that infinity can be seen for what it is, the *realization* of that possibility—a totality of all numbers.

It would be like the difference between sighting down railway tracks as you walk them, each step revealing a bit more continuation of the tracks, and viewing the entire railway from a vantage point high enough overhead.

M. Yes. That helps me to see how you're picturing things when you say that it might be that some *one* thing has better and worse formulations. Now, what about the other part?

P. Well—here you might speak of two different kinds of infinity. One which is merely potential, and one which is complete.

Only the *complete* infinity would be subject to a highly developed theory of the kind inaugurated by Cantor. And so the advantage would lie on the side of the complete infinity as opposed to the merely potential infinity.

M. Perhaps—but why hold one concept to be *more correct* just because it facilitates the development of a more complex theory?

P. Fair enough. But it would be hard to accept that there are two infinities here—investigated by distinct disciplines. I think there must be only one. Either it is the potential infinity discerned from within arithmetic, and then a whole lot of set theory either has to go or has to be reinterpreted, or it is the actual or completed infinity discerned from within set theory. And if that is the real infinite, then potential infinity is just *an approximation* to reality, and is not itself a real thing at all.

M. What you say strikes me as both important and obscure.

For my own part, I am inclined to question rather than to employ the notions of 'potential' and 'actual'. They seem to me more pictures than thoughts. As for set theory and arithmetic, well —there they are. I seek to understand them. And I am quite ready to *compare* them as well. But it is hard for me to see how there can be a better or worse here—much less a right and wrong.

But perhaps we should now set these very difficult ideas about different sorts of infinities—or different notions of the infinite—aside. For I would like to return to our main subject, the idea that something is known in arithmetic. For that is what we've really been talking about.

Is that agreeable to you?

P. Certainly.

M. So let's go back to the simplest of axioms and ask whether you know whether or not zero is its own immediate successor?

Or if you want something even simpler, do you know whether or not zero has an immediate successor? Maybe zero is like a man who has no children!

P. Well, I'm assuming that the immediate successor function is defined for the natural numbers and thus for zero, and a function is well-defined only if it has a value for each of its arguments. So there must be such a thing as the immediate successor of zero.

M. Are you proposing that you know something on the basis of some *assumption*? Can assumptions yield knowledge? Is that your idea?

P. No. I'd be ashamed to claim that. For all sorts of false things can be assumed, and other things believed on their basis. And this is no way to know something.

M. So—the appeal to the assumptions we make about functions or function symbols when we write out our axioms won't help because assumptions are insufficient for knowledge.

P. I agree.

M. Let me ask an even simpler question, one you will perhaps be able to answer with less difficulty. Tell me, do you know whether there is any such thing as zero? For I suppose that if there are any numbers at all there will be at least that one. Or, if you were to have more confidence in one than in zero, just go ahead and answer my question in relation to it.

P. What could make you think that the simpler questions are easier to answer? They often seem to be exactly the questions we have no idea of how to answer!

But perhaps we can find an answer to your latest question in set theory, for it is utterly baffling from within arithmetic what might show that there are numbers.

In fact I should have noted this from the start and then pointed out that we cannot *within* arithmetic establish that there are infinitely many numbers since we cannot *within* arithmetic establish even that there is one number. In arithmetic we merely assume all this.

But it may be different with set theory, for in set theory we can actually establish that there is an empty set, and just one.

M. But you surely can anticipate what I will now ask,

P. Yes. You'll ask how I know the things I must know in order to establish that there is an empty set. Well, the proof will be by appeal to a separation axiom, the one which says, e.g., that for any set S there is a set T the members of which are exactly the members of S which are not self-identical. Since no member of any set is not self-identical, since nothing fails to be itself, the set T will be empty. So there is an empty set. By the axiom of extensionality, every empty set will be T. Thus, there is one and only one empty set.

And now you will ask me about the extensionality axiom. You will ask me how I know that sets are identical if they have the same members. And you will note that the separation axioms yields the set T only if there are sets—at least one. And so you will ask me how I know there is at least one set.

Or you will ask me just to tell you how you might find it out on your own.

You see how good a student I am. I have learned all your questions.

M. Yes, you are getting the hang of it.

P. Now, why don't you be an even better teacher than I am a student, and teach me how I know there are sets!

M. But I do not even know that this is something you *know*. Nor do I know that you *know* there are numbers.

I of course am not concerned that you are ignorant—that you don't know something of the kind apt for knowledge. But maybe what we find in mathematics, and in set theory as well, is not knowledge, though also not mere belief or belief with a justification which falls short of knowledge.

I do not of course mean to call into question that something happens among us that we call learning arithmetic or learning set theory, and that there are experts in each field.

But it is not clear to me that learning arithmetic or set theory is acquiring information about anything.

P. What do you mean?

M. Well, do we in learning to add acquire information?

I ask myself questions like that, and sometimes, for a moment, it almost seems to me that I can see that we thereby acquire no information at all. But then, of course, most often I suppose, along with nearly everyone else, that in learning to add I learn how to acquire information about numbers, for example, that the sum of 7 and 5 is 12.

So, it is almost as if I only *wonder* whether I wonder about whether knowledge has anything to do with mathematics and set theory.

P. I think you've been reading too much of the wisdom of the obscure—if I may so refer to those Austrian sayings you're addicted to.

But, to speak more seriously, there is something decidedly odd about saying that we find out that there are numbers, or sets. On the other hand, it seems clear that there is mathematical and set-theoretical *knowledge*, and that some of this knowledge deals with the existence of numbers, for example the prime numbers. But since one could hardly know there are prime numbers greater than seven—and this is something we do know—without knowing that there are numbers, that there are numbers must also be *known*.

So, if we really don't *find out* that there are numbers, it looks as if we must eventually speak about things which we know but which we don't find out. At bottom there must be, in some curious way, knowledge without discovery—knowledge without *coming to know*.

M. How strange! Or how familiar—for I think Plato once said something similar to this. (Plato 1985, 81)

P. Even more strangely, it seems to me both that I know there are numbers and that there is no way that I could know this. For though it seems to me that I know there are numbers, it also seems clear to me that there is no way I could get hold of them so as to know that they exist. I certainly cannot *perceive* them, and I have no idea whatsoever about how to go about finding them or coming across them. And that makes me think it is impossible to know there are numbers.

So it seems to me that I *do* know there are numbers and that I can't know there are numbers!

I just don't know what to say. Perhaps I do lack knowledge here. But then I hardly feel ignorant of the existence of numbers.

M   Could it be that you both have no knowledge and yet in no way lack knowledge?

P.   Of course not! What a silly suggestion! Either I know there are numbers, or I lack that knowledge. And, similarly, either I know there are infinitely many numbers or I lack that knowledge.

It would be a contradiction to say that I both do not know and do not lack knowledge. Unless, of course, there are no numbers. But we are here assuming that there are numbers and asking only whether or how we know there are.

M.   That is a strong reply. So let me try this. Could it be that you neither have knowledge nor are ignorant because you have something in between—a kind of probable or likely belief?

It might be that you possess *evidence* that there are numbers, but not *decisive* evidence, or that your reasons for thinking there are numbers are good reasons, but less than conclusive reasons.

What do you think?

P.   It could be that way. But then the evidence or reasons would have to somehow lie outside of mathematics itself. And though I think we could explore that possibility, it at least *seems* to me that it is by thinking about numbers that I can see that there are numbers. So I would prefer to continue along the lines we have been following.

M.   Yes. Whether we can know there are numbers only by reflecting on something external to mathematics is a topic well worth pursuing. But, like you, I somehow feel that it is by thinking about numbers that we know of them, if knowledge has anything to do with it.

Some philosophers who may also have felt this way, that it is by thinking about numbers that we can know of them, have found the courage, if that is what it is, to give us a short answer to our question. They say that they know there are numbers by immediate insight or intuition.

Perhaps this is a view we should consider.

P. So they say that they know there are numbers, and when we ask them how they know, they say, 'By intuition'. Is that it?

M. It seems so.

P. Then why don't they just say that they know but don't know how they know? I think that would be more honest and thus more befitting of a philosopher.

M. I would agree, if that is all it came to. But they have more to say.

They would point out that you yourself feel you know there are numbers. But you have no answer to the question how you know this—as if the request for evidence or grounds or reasons should be met by the honest answer 'There is no evidence in this case, no grounds or reasons'. For here we seem to know something even though there is nothing which shows us that what we know is so.

After all, they will add, hasn't it often been said that knowledge by evidence must come to an end in self-evident knowledge? Or hasn't it been put this way, that things can be shown only if something is seen without anything else showing it?

Perhaps, they will urge, the truth of the matter is that we can see that there are numbers even though nothing shows us that there are numbers. They will agree that in the usual case we first conceive of something and then make an investigation to determine the truth of the matter. But it may be that in regard to the numbers no *further* investigation is needed. All we need do is to sharply and fully *conceive* of the numbers to see that they also exist.

P. Then that would be a queer fact about them. For I certainly don't have any such recognition.

M. But don't you? You say that you are convinced that you do know there are numbers, and that nothing shows this to you or could show this to you. That must mean that nothing could be *evidence* that there are numbers. So, your own view seems to be that you know there are numbers and that there *couldn't* be any evidence for this, that there *couldn't* be anything which shows this. So, it looks as if you too simply recognize the truth of the proposition that there are numbers upon actually grasping it.

P. I hadn't thought of my situation in that way. It is as if the thought that there are numbers, when I fully and clearly bring it before my mind, makes its truth evident to me.

M. So is this perhaps what we ought to think? Will you want to answer my question along these lines and say that you do know that there are numbers but that there is no answer to the question *how* you know, because nothing *shows* or even *could* show that there are numbers? Or you may perhaps say that you know that there are numbers by clearly understanding the proposition that there are numbers—that in fully grasping this proposition you 'intuit' its truth. .

How do these explanations suit you?

P. I am not sure. But I will adopt them for the sake of our discussion. Have you anything: to say about them?

M. Only this. I would ask what is required for a clear understanding of the proposition that there are numbers, or for fully grasping it.

For I suppose that you will not assert that everyone who understands this proposition will therewith assent to it—since so many philosophers have not assented to it, or have done so only with grave doubt and uncertainty.

Is that right?

P. The facts are as you state them to be. So it must be that what is required for the recognition of truth is a really clear or full understanding or grasp of the proposition.

Unless, of course, those who have doubts about the existence of numbers have them because of fanciful reasons which no person of sound and unbiased intelligence would think of, or because they somehow fear the significance which would attach to the admission that such things as numbers really exist. Since that would compel them to recognize realities beyond those of material existence or which are accessible to perception. For you know how strongly the prejudice, if I may call it that, in favor of the physical and the perceptual is common today among philosophers.

But however this may be, it does not seem that not just any understanding of the proposition that there are numbers will suffice for the recognition of its truth. So your question is worth asking.

M. Then let's seek its answer. What would go into fully understanding one of the propositions we've been considering— and let's stick with the simple one to the effect that there are numbers, zero being one of them.

P. I suddenly don't know what to say. Let me think for a moment.

I'm not sure. You say the word to yourself and just understand it. Something like that. You just form the idea.

M. But didn't we agree that some people have the idea without the knowledge? So—wasn't it our view that they perhaps didn't *fully enough* or *clearly enough* have the idea? So tell me, what goes into having it fully and clearly enough.

P. Well—it must be a particular act of mind that grasps the concept in such a way that the concept is fully and clearly present to the mind.

M. I'm not sure that it helps to say that it is a *particular* act of mind. Every act of mind is that. And I agree that it will be a case of a concept being clearly and fully present to the mind. But my question was, what goes into that?

P. How can I answer? The actions of the mind are hard to pin down.

M. Well—let me ask this then. What is it you have in mind when you speak of a concept or idea? If we got clear about that it might help us to know something about the kind of act of mind which is related to a concept or idea.

P. I think the better word is 'concept'. For a concept is an abstract object of a certain kind and not a mere psychological configuration, which is what an idea is. And it has a semantical value, since things may or may not satisfy it. So I will say that it is an abstract semantical object.

M. I suppose you will next tell me that numbers also are abstract objects, though not semantical ones. Is that so?

P. Yes.

M. So you know that one kind of abstract object exists just as soon as another abstract object is fully present to your mind. The second abstract object 'represents' the first one, and just by getting the second one fully before your mind you can see that there really are objects of the kind it represents. Something like that will be the story?

P. I think so. Do you find any errors in it?

M. None at all. I don't even think I as yet understand it. So tell me, what is an abstract object?

P. A number is an abstract object. So is a concept or a proposition. These are the kinds of things I have in mind. And these are things which exist but which have no spatial positions and are quite outside of time.

M. I doubt if any sense attaches to your remarks.

P. How so? Talk in terms of abstract objects generally passes without objection, and it is a well-known thesis defended by many philosophers that there are abstract objects and that numbers and concepts are among them. I don't mean that philosophers never disagree about whether there are any abstract objects. They even disagree to some extent about how to best understand the term. But no one supposes that to say 'Numbers are abstract objects' is to fail to make reasonably good sense.

M. Well, it is clear that the term gets used, and that philosophers argue in certain ways in their use of it. But it is not so clear that anything actually gets said in the use of this word. At least that's not so clear to me.

If someone tells me that four is *abstract*, then I'm just not sure what, if anything, they're telling me. Oh, they may say that part of what they mean is that four is not in time. But then I'm not sure what that means. And when they say that four is not in space, well, that's also unclear to me. They may say that four *cannot* cause anything to happen. But do I understand that? Familiar words are put together in unfamiliar ways, and I think I just don't get it, and suspect that those who think they do get it are under an illusion.

P. You say you don't get it. But don't you? If I tell you that beauty isn't in space, won't you understand that?

M. Well—I go into a room and look at a painting hanging on the wall. It is quite beautiful. Then you tell me that though the painting is here in the room, its beauty isn't. Do I understand that? Not yet I don't! And if you tell me that though there are three tomatoes in the basket, their number isn't in the basket, well I don't feel I understand that remark any more than the remark about beauty. Not at all.

P. But if someone said that beauty was hanging there on the wall, or that the number three was in the basket, wouldn't you deny that? And wouldn't that mean that you saw that beauty was not there in the room, and that three was not there in the basket? And when you say that beauty or the number three are not in space you're just summing up these obvious points along with all similar ones. For no matter where you pointed, neither beauty nor the number three would be there.

M. But you assume that I would deny that beauty was hanging there on the wall, or that three was in the basket. I certainly wouldn't *accept* the words, but my refusal to go along with them would not be a denial, but an expression of *puzzlement*.

P. I can see how it may be puzzling to you whether to say that beauty is in space, or to say it is not in space. But what beyond that is puzzling to you?

M. The form of words 'The beauty of that painting is in this room' makes no clear sense to me. My puzzlement has to do with what is said—not with whether what is said is correct or incorrect.

I can't deny what I don't understand. That I also don't accept what I don't understand is no reason for thinking that, as it were, I believe the opposite. Here the opposite is just the negation, and I no more understand it than I understand the sentence of which it is the negation.

It isn't that I see no way to make sense of this talk. One might stipulate that the sentence 'Beauty is in this room' is to mean that something beautiful is in the room. Well, then, OK. But I doubt that philosophers will say that this is what they mean, for if this were what they mean, then it would be pretty obviously true that beauty is in any room in which there is a beautiful painting.

But as things stand, I doubt that any sense attaches to this philosophical talk in terms of abstract objects.

P. This is crazy. I don't know how many times I've heard you talk in terms of abstract objects. I won't say that I always agree with you when you use this term, but you certainly show that you have grasped the general idea perfectly well. Some of your arguments have even struck me as ingenious.

M. Look. I read the same stuff that we all read, and I get the hang of it. I know the moves—not all of them, but enough of them. But it is all so vague and indeterminate. We say these wonderfully professional sounding things. We say, as if it were as clear as clear can be, that abstract objects are causally inert since they are not part of space-time. We go on like that, and whole books may get written which weave talk in terms of this term through their pages.

But it may be that all our 'understanding' here is nothing but a feeling of understanding—a certain familiarity with a way of speaking, and a knack for engaging in it.

P. You know, you have an all too easy way of dealing with philosophy. You claim you don't understand what is said, and insinuate that this may not be a deficiency in you, but a lack in the words of the philosophers.

But among the philosophers who thought we can speak of beauty and *correctly* say that beauty is not in space was Plato. And it isn't much of a reply to his philosophy to say, in effect, that you don't understand it because it makes no sense.

M. If Plato said it, or if Frege said it, then there is bound to be *something* right in saying it. I'm not denying that. In what may be a similar way there is *something* right to saying that numbers aren't colored, or that sighs are weightless. But what is it that we are driving at when we use such turns of speech?

You suggested that beauty is not in this or that room. And then, as if you were generalizing, said it is not in space. But you don't say that beauty is not in this room because you didn't find it there. You have no idea of what it would be like to search the room for beauty — as you might search it for a mouse.

Might it be that in saying that beauty is not in space we are pointing out that finding beauty is not at all like finding an object at some place? There would be a point in that.

P. Yes. And just as there is such a thing as finding beauty, there also is such a thing as searching for it. I might search for beauty in a forest, or in a room. And find it too!

M. For example, you might find a place of beauty in the forest, a glade perhaps. And now we could say that is finding beauty in space. And if we spoke in this way, well then it would often be true that beauty is in space.

But, still, beauty is not in the forest in the way the glade is. And that could be said as well.

P. Stop right there! Haven't you just made sense of 'Beauty is not in the forest'? You just used those very words. Perhaps what you just said is exactly what Plato meant!

M. But you misunderstand the 'not'. What I said could be put this way, that it is not the case that beauty is in the forest in the way the glade is. And that does not imply that beauty is not in the forest. If I say that she does not love her child in the way she loves her mother, I do *not* at all imply that she does not love her child. Do you see what I mean?

P. Yes. I went too fast.

M. I think that what I wanted to indicate by my turn of phrase was a way in which to understand the words 'Beauty is in the forest' — namely, as saying that there is in the forest something of beauty. Understood in this way, it is right to say that there is beauty in the forest. To then add 'but *not* in the way the thing which is beautiful is in the forest' makes it clear that here a term is given a special use.

The glade is in the forest by being one of its parts, whereas beauty is in the forest by some part of the forest being beautiful.

We might also say that searching for beauty in the forest is not like searching for a stream or a glade, even if you find beauty when you find the glade. It is more like searching for pity in a human face.

As to whether Plato had anything like this in mind, I am not at all certain.

P. Still, following your lead one might give to 'Three is in the basket' the sense of the sentence 'Three of something are in the basket', eggs, for example. And then one would add that three is not in the basket in the sense in which the things of which there are three in the basket are in the basket.

M. Yes. One might adopt this strange way of speaking. But it provides a sense for 'Three is in the basket' on which it may well be true that three is in the basket—for example, if three eggs are in there. So this does not get us to talking about three not being in the basket.

P. Agreed. But then we add that three itself is not in the basket—the *number* isn't there, only the things with that number are there.

M. But how are we to understand this phrase 'Three itself isn't there'?

We might take it to come to this: Remember, when we say that three is in the basket we mean only that three of something is in the basket.

That will be entirely intelligible and in order. But it hardly carries the weight, if I may so put it, that the *philosopher's* 'Three itself is not in the basket' is felt to possess.

Just as it stands, without explanation, the form of words 'Three itself is in the basket' is not one I understand. Of course, I understand its *words* and even see what sort of *grammar* it has— for grammatically it is much like 'Bill himself is in the room'! So there is a sense of familiarity to the sentence 'Three itself is in the basket'. But that there is a sense to the sentence, well, that is not so clear. That a sentence has a familiar sound to us is not the same as our understanding it.

P. I feel a certain impatience at your fastidiousness about sense. We know well enough how to talk in terms of abstract objects. You admitted as much yourself. So let's drop your refined worries and just get on with our discussion of knowledge of the existence of numbers.

M. OK. But remember that I shall now be freely engaging in a kind of talk in which I have no confidence at all.

Let's first take a look at the view we were considering. You said that numbers are abstract objects that we know to exist as soon as we have concepts of them clearly and fully present to the mind. Then you said that these concepts also are abstract objects. Is that right?

P. Yes.

M. So I can conclude that an abstract object can be clearly and fully present to the mind. But in that case, why bring in the concepts by *means of which* numbers are conceived? Just let the numbers themselves be clearly and fully present to the mind.

P. I see your point. On this theory some sort of abstract objects will themselves have to be present to the mind, without any intermediary, as it were.

M. Further, your theory explains our knowledge of some abstract objects by reference to other abstract objects, the concepts. So the very question that arose about the numbers arises about concepts as well: How do we know they exist? Will you say that we know they exist by bringing clearly and fully before the mind the concept of *concepts*? But how do we then know there is *that* concept?

P. I think the answer to this question is that we know *immediately* that there are concepts because they are *directly present* to the mind.

M. But, then, to go back to the first point, you might as well say that the numbers are *immediately* known by being *directly present* to the mind. I really think that bringing in concepts may be an unnecessary step.

P. Then let this be my view, I know of the existence of numbers because they are *themselves* directly present to my mind.

M. That may be the best line to take. But remember the problem you felt about the numbers, that they could not be perceived? That was part of what at first made you feel you couldn't know about them.

Well, we now have the theory that the numbers are abstract objects that exist but not in time or space. So, as philosophers are quick to point out, they are, on this account, causally inefficacious. But then they cannot interact with or in any way effect our sensory organs or neural processes.

So, what in the world can be meant by this talk of 'being present to the mind'? It looks as if the 'mind' must here be understood to be as non-physical as the abstract objects themselves. But are you ready to go for a view of our knowledge of numbers which involves saying that they are known by some kind of connection between abstract entities and a non-physical mental substance? Or something of that sort?

P. I admit that I find it hard to believe this story.

M. I would even be inclined to say that it may be *impossible* to believe this story, since it may be just more mere talk—sound without sense.

P. I don't know quite what to say. It seems to be very clear that we know a lot about the numbers. It also seems to be very clear that numbers aren't any kind of physically existent objects. And if these points are both correct, then there must be a way in which we human beings can come to know about non-physical objects.

M. But haven't philosophers persuasively argued that it is precisely the *impossibility* of human beings having knowledge of what they cannot causally interact with which shows us that it *cannot* be both that numbers are non-physical objects and that we know of them?

P. Yes. And this view is persuasive. But then other philosophers have questioned it, and have employed subtle arguments of their own to show that the alleged impossibility is far from established. So I again feel lost at sea!

M. Our conversation has descended from the confident assertion that there are infinitely many numbers, to an unresolved bewilderment about how we can know there are any numbers at all.

P. Yes, and in the course of our conversation we also fell into confusion about what the infinity of the natural numbers is, or about how it is rightly to be understood—either arithmetically or set-theoretically, either as a potential or as an actual infinity. And it wasn't even clear to us that there were two really different ways of viewing infinity even here! At every turn we seem only to realize more fully how little is clear to us in our thinking about mathematics.

M. I agree. And for me one of the most striking features of our conversation is, if I may so put it, how one thing leads to another. It fits in with my idea that it is impossible to deal with one problem in philosophy without being inexorably drawn into all the problems of philosophy.

   Not that we actually were lead to consider all of them, for that would not be possible, but that we could see again and again how, if we were to follow up on some idea which naturally suggested itself, we would be led from topic to topic.

P. Philosophy is like exploration. You can never know in advance the particular lay of the land to which you will next advance, and can never tell in advance whether a particular path you follow into a new territory will actually make it accessible to you. And as you track back into provinces already visited, they often look different—perhaps because of the way the lands newly seen cause you to see the old lands in a new way, or simply because you now approach the familiar lands from an unfamiliar angle. In a way it is frustrating—for the unanticipatable element in exploration precludes certainty even about the lay of the land already covered. But it has its own excitement—the prospect of yet again seeing something different and thereby seeing differently even what has already been seen.

M. Yes—your words mean something to me. But I really wouldn't mind a *bit* of certainty now and then!

CHAPTER 5
A Conversation of Five

This is a dialogue in which our five characters—C, M, P, N and A—get together. Various issues in the philosophy of mathematics are discussed: numbers as abstract objects, our knowledge of numbers as abstract objects, a proof as showing a mathematical statement to be true as opposed to the statement being true in virtue of having a proof, the problematic nature of the notion of the truth-condition of a mathematical statement, formalism and classical logic.

The Dialogue

C. Frege asked what numbers are, but not whether they are. I think he just took the point for granted. But it certainly isn't obvious to me that there are numbers!

P. Well—there are existence proofs in mathematics, and lots of very obvious ones—for example the proof that there are prime numbers greater than 3. And if there are such numbers, then there certainly are *numbers*.

   So mathematics itself settles the issue of the *existence* of numbers.

C. Doesn't mathematics itself equally much settle issues about what numbers are?

P. Mathematics certainly tells us a lot about the numbers. But mathematics does not itself settle what they *are*. And that was what interested Frege. What are they? Concepts? Objects? And if objects, objects of what kind?

   His conclusion was that they are objects, but of a quite special kind. And he sharply distinguished them from the signs we use to signify them as well as from such ideas as we have of them.

So this is how I look at it: The existence of numbers is something mathematics itself speaks to. But Frege's questions were ones on which *mathematics* is silent. Work *within* mathematics doesn't even address these questions.

Frege's question is a philosophical question.

A. I see a point in your remarks. And for me, as for Frege, the *interesting* questions concern the nature of number. I often wonder what they are, and sometimes seem to feel that they are symbols of some kind, perhaps signs, and perhaps ideas.

P. Frege said they were objects, but not signs nor ideas.

A. But if they are neither signs nor ideas, then *what* are they? Call them objects if you will, but then what kinds of objects are they?

P. According to Frege, they are objects of a special type. They are objects outside time and space—objects which do not exist anywhere or at any time.

A. As far as I can see, that's as much as to say that they don't exist at all. So, if Frege was right in what he said numbers were, then there aren't any!

P. Not at all. To say that numbers are neither in time nor in space is to *characterize* them, not to deny them. They are existent objects of a kind *quite different* from those associated with either mind or matter. To give them a familiar name, they are *abstract* objects.

C. Wonderful! Theologians tell us that they characterize God when they say that God exists neither in time nor in space. And then, as if it would clear everything up, they tell us that God is not a material being. Some characterization!

P. I agree that to say that an object is *not* in space and is *not* in time doesn't say much about what it *is*. But to say that an object is abstract is not like saying it is immaterial—it is not merely to say what it isn't, but to say something of what it *is*.

C. Of course not! To call something an abstract object is not to characterize *it* at all. It is only to say something about our *conception* of it. It is to say it is an object that we conceive of by abstraction.

It's not that I mind saying that—though I don't know that I really understand such talk—but it just doesn't tell us anything about what *they* are. So we are left with the claim that they *are* at no place or time. And that is just to say that they aren't at all. I think A was right about that.

N. For my part I would not quarrel over whether the term characterizes numbers or our idea of numbers.

For me the main point is that even though it is unclear to me just how we form our ideas of the numbers, there certainly is no problem with the idea—however formed—of objects which really exist but do so neither temporally nor spatially.

You two, A and C, seem to think it is impossible for something to exist but not exist either in time or space. But I can see no bar to their being such objects. There is nothing *impossible* in the idea of such an object—and so, there even are possible worlds including such objects.

C. So say you can't see anything impossible in the idea of an object neither in time nor in space. All that comes to is this: For *all you know* there are such objects. And that comes only to this: *You* see no impossibility in the idea of such objects. But from that it does not follow that there could be such objects.

N. No, it is not merely that I *cannot see* anything impossible here. Rather, I *see* that such objects *are* possible, for I see that it is logically possible that there are such objects.

After all, to say that numbers do not exist in time and do not exist in space is not to say flat out that they do not exist, much less to imply that they could not exist.

M. Is it your idea that whatever is logically possible is *possible*?

N. Would you suggest that something logically possible might be *impossible*? I think you will not want to go that far.

M. Well, it is not *logically* impossible that some bachelors are married. But that is not *possible*.

N. How can you say that it is logically possible for there to be married bachelors? After all, that reduces to a logical impossibility by definition. And I count as logically impossible whatever reduces to a logical impossibility by definitions.

M. It also is not *logically* impossible for some surface to be entirely red and entirely blue. But it still is impossible, and in this case there are no definitions whereby we can reduce this impossibility to a logical impossibility.

N. This really is tiresome. There is a narrow and a wide sense of logical impossibility, and I *of course* meant logical impossibility in the wide sense.

M. I have an idea of what logical impossibility in the narrow sense comes to. That's the sense that gets laid out in logic texts. I would say of this sense that it is the *ordinary and familiar* sense of logical impossibility.

The broad sense is just undefined. You don't *want* to say that such and such is impossible though logically possible, so you say that it is logically impossible *in the broad sense.*

I would say that the so-called *broad* sense of logical impossibility is not a sense of *logical* impossibility in any reasonably well understood way.

If I ask you in what way it is broadly *logically* impossible for a surface to be entirely red and entirely green, what can you say?

N. I would say that it is a point about the *logic* of color.

M. And what is this logic of color? Isn't it just the various *necessities* and *impossibilities* concerning colors and their relations?

N. Yes, just that. We see that it is necessary that pink is a shade of red, that nothing is both entirely blue and entirely red, and the like.

M. Now tell me, what is *logical* about these necessities?

N. I find I have nothing to say on this, or at least nothing at all definite.

So I will not say that I see that it is logically possible for a thing to exist but not in time or in space and *thus* possible, but shall instead simply say that I see that this is *possible.*

M. Good. And now, should we wish to deny that this is possible, we will not feel that we are going against *logic*—as we would were we to claim that it is possible that married men are not married!

N. I have heard that you like to make small points. And now I see that you do. And I suppose we could have quite a discussion about possibility. But that really is not very important to me. I mentioned the point only to make it clear that I wasn't concerned with the issue of whether or not numbers exist. My problem lies in an entirely different direction. For if numbers do not exist in space and do not exist in time, then we cannot have any commerce with them. It is Plato's problem about knowing the forms all over again.

So, if numbers exist neither in time nor in space, then so far as we can *know*—so far as we *could* know—there aren't any!

And this is not because we fall short of knowledge due to some slim possibility of error, but because we neither have nor could have *any* reason for thinking that there are any such things. That there are numbers would not be even a reasonable or plausible *conjecture*.

P. This problem has its solutions, just as did Plato's. For example, when you say we could have no commerce with them, you're thinking of the impossibility of perceiving them. You think to yourself: If I can in no way perceive them, then I cannot know of them. So you assume that our power of thought is tightly tied to perception, and this assumption may be in error.

Indeed, we can reverse your line of argument and say that *since* we know of the numbers *and* they exist neither in time nor in space, our power of thought is *not* so tightly tied to perception as you assume.

What we *can* know cannot be less than what we *do* know. And we do have knowledge of the numbers. So, since they exist neither in space nor time, the possibility of knowledge is not subject to the conditions to which you would tie them.

N. I of course agree that what we can know includes at least what we do know. But we *couldn't* know of any such things as Frege says numbers are. And so we *don't*.

P. So, you agree that there may be numbers—things which exist but exist neither in time nor space. But you hold that even if there are such things, *we* cannot know that there are such things.

   Is that something you *know*?

N. I think I do.

P. Well, suppose you do. Then you do know something about such things. But in that case your claim that we cannot know anything about them is false. And so, that we cannot know anything about them is something we *don't* know. Nor *could* we know it!

N. I think anyone can *feel* the sophistry in what you've just said.

P. Perhaps, but do you find any error in my reasoning?

N. Well, do you find any error in mine? Here is how I reason: If something exists neither in time nor in space, then we cannot know anything about it. I think we know this much. Then I say that if there are numbers, then they exist neither in time nor in space. I think we may know this too. And so we can also know this: If numbers exist, then we cannot know anything about them. Now suppose that numbers do not exist. Then, certainly, we can know nothing about them. So, if numbers either exist or do not exist, we can know nothing about them. But we know they either exist or do not exist. So, we know that we can know nothing about numbers.

   This reasoning seems to me very cogent.

P. We can put your conclusion as follows: Numbers are such that we can know nothing about them.

   That conclusion most clearly is about *numbers*, and so if it expresses something you *know*, then you do know something about numbers.

N. By the same reasoning you could say that we have knowledge of centaurs. Consider the conclusion: Centaurs are such that they do not exist. This is about *centaurs*, and is something we know. So we do know something about centaurs.

   But of course we *don't*—for there aren't any.

P. Do you really think we cannot have knowledge of what doesn't exist?

N. Of course we can't. What could be plainer than that?

P. But you do not deny that we know that there are no centaurs.

N. We most certainly know that.

P. So, in *one* sense we can have knowledge of what doesn't exist—for we know that centaurs don't exist—though in *another* sense we cannot have any knowledge of what doesn't exist—for there are no centaurs for us to know anything about them.

N. So it seems.

P. And your idea is that in the sense in which we cannot have any knowledge of what doesn't exist, we also cannot have any knowledge of numbers since numbers exist neither in time nor in space. But in the sense in which we can have knowledge of what doesn't exist, we can have knowledge of numbers even if they exist neither in time nor in space.

N. That seems right. But I feel confused about these two senses? Is the word 'knowledge' somehow ambiguous?

P. I would not so much say that as say that it is a matter of there being two kinds of knowing. There is the kind of knowing that doesn't require any kind of contact or 'commerce' with particular things, and the kind of knowing which does.

Your knowledge, if you actually have it, that we can know nothing about numbers, is knowledge of a kind which does not require any kind of contact with particular numbers. So, if you are right, then you do know something about numbers—you have *that* kind of knowledge of numbers. And *what* you know is that you lack the other kind of knowledge of numbers.

N. When you put it that way, I am reminded of terms I have heard from philosophers when they talk about belief. They say that some of our beliefs are *de dicto*, and that others are *de re*.

Could you say that I know *de dicto* that we cannot know numbers *de re*?

P. Perhaps. But you tell me. Was it your idea that you know *that* even if there are numbers, since they are neither in time nor in space, there are no numbers *of* which you know anything?

N. That surely is part of my idea. For the rest I would say that even if there are numbers we couldn't know *that* there are numbers since to know *that* there are numbers there must be numbers *of* which we know that they exist. What I mean is this, that knowing that there are things of a certain kind is always knowledge of the kind which requires contact or 'commerce' with things of that kind.

   So I might put it this way: *de dicto* knowledge that there are numbers requires *de re* knowledge of numbers.

P. Yes, though that shows that *de dicto* knowledge is not exactly the same thing as knowledge that requires no contact or 'commerce' with particular things.

N. I see that. But the important thing is that you have showed me a way out of the paradox of claiming that I know that we know nothing about numbers.

   The way I would now want to put my view is this: since numbers are things of a kind which exist neither in space nor in time, then either there are no such things—in which case we can have no knowledge *of* them—or there are such things—in which case we again can have no knowledge *of* them since we could have no commerce with them. So what I claim is to know *that* we have and can have no knowledge *of* numbers.

   And my main point is this: since numbers are things of a kind that exist neither in space nor in time, we can have no knowledge of numbers.

P. So here is how things stand, leaving A and M out of it. C thinks there are no numbers, and N thinks we can have no knowledge of numbers. So you agree on *this*, that what ordinarily goes by the name of mathematical knowledge isn't *knowledge* at all, but is either false belief or utterly unjustified belief.

   In that case, mathematical knowledge refutes you both.

C. N denies that there is such knowledge, I don't. But I think I know why you think I do. It is because I deny that there are any objects answering to mathematical signs and you think that mathematical knowledge includes the knowledge that such objects exist.

But to interpret mathematics as asserting the existence of entities of any kind whatsoever—whether mathematical signs or objects answering to such signs—is to misinterpret it.

It's an error to say that the proofs of mathematics establish *existence*. They establish only what we call *mathematical* existence. That mathematical existence involves the existence of objects answering to the signs of mathematics is *quite another point*. To say so is a piece of philosophy, not a truism of mathematics.

It is an issue—just like the issue of the nature of numbers—on which mathematics itself is silent. And that is what I should have said at the start, when you first suggested that mathematics itself settles the existence of numbers.

And so I still think that Frege should have first asked whether there are any numbers.

P. And how would you answer that question?

C. I hold that the *right* way to regard mathematics is as a discipline which deals with no existent objects whatsoever, and which doesn't intend to. Mathematics is precisely what it looks to be: a system of notation within which we carry out proofs. What gets established in establishing 'mathematical existence' is just a certain type of mathematical sentence, and its truth is nothing more than its provability.

And here the key point is this: In mathematics proof establishes nothing beyond itself.

Proof inside mathematics is completely different from proof outside mathematics. In the non-mathematical cases, proof establishes something more—truth. In mathematics, being true *consists in* having a proof.

P. That seems quite wrong. A proof always shows that something is true, but does not itself make for the truth of that which it shows to be true. A proof can show that some statement is true, but the statement is not true in virtue of its having a proof, but in virtue of things being as it says they are.

C. I insist that this holds only in non-mathematical cases. I agree that what ordinarily *establishes* a sentence is not what *makes it true*. What makes for the truth of a sentence like ' There are whales' is how the world is, not the sequence of sentences we set forth as establishing this sentence. But this familiar distinction lacks application within mathematics.

P.   Confidently asserted! But let's *look* at what you've said. You used the phrase 'a system of notation'. But a notation for *what*? You didn't answer that question. You didn't even ask it! But it is there for the asking, and its answer plainly is this: Mathematics is a notation for constructing statements about *numbers*. That is a virtual truism.

And one thing which shows that the notation so serves is the fact that its sentences divide into those which are true, and the rest, which are false, not merely those which have proofs and those which have disproofs.

In mathematics as elsewhere, truth is one thing and provability is another. Mathematical sentences, like other sentences, have truth-conditions. And their truth consists in those conditions being satisfied.

The truth-condition of 'Snow is white' is that snow is white. So also, the truth-condition of '7 + 5 = 12' is that 5 + 7 = 12. The physical facts make for the truth or falsity of sentences about snow, and the mathematical facts make for the truth or falsity of sentences about 12.

M.   But then the truth-condition of '7 + 6 = 12' would be that 7 + 6 = 12, i.e., that 13 = 12. What is the sense in saying that it is a condition of the truth of anything that 13 = 12? That really is not clear to me.

Could you say again what this truth-condition of which you speak is supposed to be?

P.   Try this: A truth-condition is that the grasp of which constitutes the understanding of a sentence.

M.   But I suppose I *understand* the sentence '7 + 6 = 12'. But what sense does it make to say that I grasp that 7 + 6 = 12 or that 13 = 12?

P.   If that bothers you, put it this way: To understand a sentence is to *know* the condition which, were it to obtain, would render that sentence true.

M. Do I then *know* the condition which, were it to obtain, would render '7 + 6 = 12' true? But there is no such condition!

P. Of course there is; it is precisely the condition that 7 + 6 = 12. Since you understand the sentence, you know that condition. The fact that it *couldn't* obtain doesn't mean that you don't or can't know it.

M. This assumes that in mathematics understanding is a matter of knowing or grasping this or that condition. But is that a correct appraisal of understanding for the sentences of mathematics?

   I understand a contradiction, and it makes no sense to speak of the conditions under which it is true.

P. Doesn't it then *lack* a condition of truth?

M. Perhaps, but it is not nonsense like 'Twas brillig and the slithy toves did gyre and gimble in the wabe'. In the course of an indirect proof we do not produce a piece of nonsense.

P. I agree. But can't we then say that it has an *unsatisfiable* condition of truth, or simply say that it is associated with a condition that *cannot* be satisfied.

M. That is not so clear to me. I am not sure that I make any sense of the phrase 'the condition of something being and not being a whale'.

P. Look, the sentence 'There are whales' is associated with a certain condition—that condition obtains or not. The same holds for 'There are whales which are not whales'—it too is associated with a particular condition. One that does *not* obtain because it *cannot* obtain. But it is not as if there were no such condition.

M. You say your words with force and confidence. But I really am not sure that I understand them to say anything.

   I understand the sentence 'There are whales which are not whales'. But do I *grasp* or *know* the condition of there being whales that are not whales? That continues to seem very unnatural to me.

P. Why are you concerned about what is *natural*? The question is whether you do or don't grasp that condition.

M. But what is unclear to me here is such phrases as 'the condition that there are whales that are not whales' or the phrase 'the condition that 12 = 13'. It is not clear to me that such phrases make any sense.

C. Notice that such worries as M's do not attach to my view, which, to repeat myself, is that while we certainly *speak* of mathematical assertions as true or false, such talk is just talk of what has a proof or a disproof. In mathematics, truth and provability come to the same. There is nothing in virtue of which a mathematical proposition is true save its proof.

P. Then you really do think that in mathematics truth coincides with provability! Well, with that you finally say something definite enough to be definitely false. Or haven't you heard of the Gödel result?

C. Look, setting subtleties aside, what Gödel showed is that if you've specified the axioms in a way which enables you to tell whether or not a sentence is an axiom, and if those axioms provide a derivation for each elementary equality and inequality in the language of arithmetic, and if you've specified the rules of inference in a way which enables you to tell whether or not a sequence of sentences is a derivation in accord with those rules, then even if the rules are complete there will be sentences which are not decided by those axioms and rules so long as those axioms are consistent.

Since you specify axioms only in terms of the marks which make up the language and since the rules of inference refer only to such marks, my answer to the objection that Gödel somehow showed it was a mistake to identify provability with truth is that what he showed concerned only marks and methods of putting them together and sequencing them. What he showed is that if you specify axioms and derivations in an *effective* way and have axioms that provide a derivation for each elementary equality and inequality in the language of arithmetic, then there will be sentences which cannot be derived and whose negations also cannot be derived. The notions of truth and falsity don't enter into it *at all*—neither by way of the content of the theorem itself, nor by way of its proof.

In particular, that truth and falsity in mathematics go beyond proof and disproof is no part of what he *proves*.

P. Strictly speaking, that is correct.

C. Don't begrudge the point.

P. I won't. The theorem and its proof are as you say. And this probably is worth noting in a pedantic sort of way.

But only in a pedantic sort of way.

For the undecidable sentences that Gödel identified and showed to be such are universal quantifications *all* instances of which are and can be shown to be provable. So *they* are all true. And in that case, the quantification is *true even though it is not provable*.

This simple consideration, though admittedly external to Gödel's theorem and its proof, cinches the case for a distinction between proof and truth. Everyone who thinks about the proof sees this, and that is why there is universal agreement that Gödel showed that there is a distinction between truth and provability.

C. Not if you count me. Let me ask a question. Is the truth of a universal quantification simply a matter of the truth of all its instances? Mightn't a universal quantification have only true instances and yet not be true?

P. Of course, but, if I may answer your objection in advance, not in elementary number theory. For here each element over which the variables range is named — for the variables range over exactly the natural numbers, and each of these is denoted by some numeral in the language of arithmetic.

C. So, the truth of a universal quantification does not *in general* consist in the circumstance that each of its instances is true.

P. I agree.

C. But in *this* case the truth of the instances is enough for the truth of the quantification. Is that what you're telling me?

P. Yes.

C. And the *reason* is this, that in arithmetic the variables range over the natural numbers and each of these is named by some numeral?

P. Of course.

C. Of course, of course. I just wanted to make it *perfectly* clear what a *dreadful* argument you're giving against *my* view.

I say that since there are no numbers, there is in arithmetic only notation, proof, and disproof, in which case truth and falsity come to proof and disproof. I agree that the Gödel sentences are quantifications all instances of which are provable. So I agree that all those *instances* are true. You then reply that this is sufficient for the truth of the quantification *since* there is a name for each *number*. But I was asserting that there are no numbers.

So, for you to reply in this way is not to construct a case *against* my assertion, but simply to *deny* it.

You certainly do *that*. But can you *refute* it? Or even make it appear implausible *on* independent grounds.

P. Yes, you did say that *since* there are no numbers, truth and falsity come to provability and disprovability. And so, since my reply assumes that there are numbers it begs the question against your view.

I agree. My objection was not well formulated. It *was* a dreadful objection.

So it comes to this: I say that there are numbers, and you say that there are no numbers, and the Gödel proof is, as it were, neutral between our two views.

C. Yes.

P. And you say that there is no notion of truth applicable to mathematics save one that rests on the notion of proof, and the Gödel proof does not speak against this.

C. Yes.

A. But I don't see how that can be. If the instances are all *true* then how could the quantification fail to be true?

M. Look—what C has said is that in mathematics truth *comes to* provability. As he put it earlier, in mathematics a proof proves nothing beyond itself.

What you're thinking is that the provability of each instance of the Gödel sentence *shows* that it is true. But C's view is that the proof shows no such thing. There is just the proof. Each instance of the Gödel sentence occurs as the last sentence in a derivation from the axioms, and the Gödel sentence itself *doesn't*—nor does its negation. And that's the end of the matter.

You keep thinking that there is a puzzle as to how in the world those instances could be *true* and the generalization *not* be true. But the notion of truth you're implicitly working with here doesn't fit mathematics.

P. You *say* that, but what *shows* it?

M. I am not at all sure how to show it. I'll leave that to C. And in fact I am really not sure that what I've just said is at all correct— though it somehow *sounds* right to me. But even if I can't show anything, I may be able to better explain myself.

In the case of an ordinary universal quantification, we do not think that what makes for its truth is the truth of its instances. I believe you even agreed to this when you said that the truth of a universal quantification doesn't *consist* in the truth of its instances, even if the truth of those instances is enough for the truth of the quantification. Rather, we think that what makes for the truth of its instances is what makes for *its* truth when the truth of those instances is enough for the truth of the quantification.

P. I don't understand this. Could you give an example?

M. Sure. Consider the quantification 'Everyone is this room is speaks English'. We five are the only people in this room. So the instances are 'C speaks English', 'P speaks English', 'M speaks English', 'A speaks English' and 'N speaks English'. Each of these instances is true, and the truth of these instances is enough for the truth of the quantification. What makes each instance true is that the person it names speaks English. And what makes the quantification true is that each of these persons speaks English and that they alone are in the room. So, what makes for the truth of the quantification is how things are with the people in the room. The quantification, like its instances, is true in virtue of how things are.

P. I think I see that. And what you're saying is that the quantification says how things are every bit as much as does its instances. Even if the truth of those sentences should happen to be enough for the truth of the quantification, the quantification still owes its truth to how things are.

M. Yes. That is exactly what I have in mind.

And so, when we say 'Since all the instances are true, the generalization is true' we aren't saying that the generalization is true in virtue of the truth of its instances, but is true in virtue of *how things are*. And *that* is what the instances also are true in virtue of.

So when you ask how the instances could be true and the generalization not be true, you're thinking that the instances are true in virtue of how things are and that the proof *shows* this.

P. So I might write down the sentences 'A speaks English' and the rest together with 'Only A and N and M and C and P are in the room' and then infer 'So, everyone in the room speaks English' and this would be a *proof*, but the conclusion would not be true in virtue of having a proof, but in virtue of how things are with the people in the room.

M. Yes. But if there is nothing more to the truth of the instances than *their* having proofs, then there is nothing in virtue of which they are true. And so, the proofs of the instances show nothing about how things are in virtue of which the generalization might be true.

To say that the instances are true is just to say that they terminate derivations. It is not that they are true in virtue of how things are, so that how things are could make for the truth of the quantification even if it did not terminate any derivation.

P. So if we had the instances and could say that they were true in virtue of how things are, then the quantification might also be true *in virtue of how things are*. But if the instances are not true in virtue of how things are, then there is nothing to make for the truth of the quantification except what makes for the truth of the instances—so that if there is no proof of the quantification it falls short of truth in the only way that it could be true. For there is nothing outside of its place in the system, as the termination of a derivation belonging to the system, to make for its truth.

M. I think that says it better than I could have said it. And although I am not sure that any of it is right, it is the sort of thing I had in mind. And I think it is what C was moved by.

P. So in a way our disagreements might be put as disagreements about proof. I think a proof shows something—namely that the mathematical sentence is true. And you two think that a proof shows nothing at all!

   Actually, I rather like that way of putting our disagreement.

C. I can see why you would. And I'll tell you, it bothers me. For although I have felt that in mathematics truth *is* provability, it somehow does not feel at all right to say that in mathematics a proof shows nothing at all. As you and N were talking I felt that just the right sort of thing was being said. But now I'm not so sure.

   To be honest about it, I think that a proof does show something—for example, a proof terminating with '5 + 7 = 12' shows that 5 + 7 = 12.

M. Of course. To say what a proof shows we simply repeat its last line. But *what* is shown?

   Not nothing. The proof is not useless. You said earlier that it does not serve to establish *truth*. So then you should ask yourself *what does a proof do?*

   I mean, of course, what does a proof do *in mathematics*.

C. I sometimes have thought that all that is going on is *showing the deductive linkages*. What we learn in mathematics is what follows from what.

M. I agree that we can take an interest in seeing whether or not this or that sentence follows from such and such other sentences. But is this the insight we are afforded by mathematics?

C. What else could it be? If proof in mathematics is not a matter of establishing truth, then all proof can show us is what follows from what!

M. Would you call it mathematics if I gave you an uninterpreted logistic system and asked whether a certain form in that system could be obtained from certain other forms by application of certain rules?

C. In a way. I might regard that as a mathematical investigation of the system—roughly as chess theory is a mathematical investigation of the game of chess.

M. Yes. But the interest you take in chess theory is very different from the interest you have in playing the game. We do not play the game to *find out*, for example, which board positions can be arrived at from which other board positions.

We achieve *something* by carrying out proofs in mathematics. But we do not carry out proofs to *find out* which sentences follow from which sentences.

C. Are you asking how we *use* mathematical results? If so, then one answer is that we use them to get further results. We prove something, and use it to prove something else.

Mathematics has a life of its own. If some theorem has an application outside of mathematics, well that is neither here nor there for *mathematics*.

M. Yes. I see that. But suppose that the whole of mathematics lacked application—in every way. Then what kind of life would it be?

C. A strange way to put a question! But I will answer: A very formal life. Like the life of the royal family now that it no longer rules. It now only obeys rules.

M. Rather like a game?

C. Exactly. The game of mathematics. Here is a conception of mathematics on which mathematics would fit views like those of the formalists or the deductivists.

P. A formal game, and a pointless game. But that is not *mathematics*. You both draw away from the obvious and natural point of view—that in mathematics we arrive at truth about numbers.

Our mathematics is not just a formal game, an artful manipulation of symbols devoid of meaning. The sentences of mathematics make statements that are or are not true, and the task of proof is the task of certifying truth.

M. The word 'certifying' surprises me. Why did you choose it?

P. My thought was that perhaps we do not *find things out by proof*. We may have other ways of *coming to know*. But that proof *shows* that what we have taken to be true is true.

I have heard of an Indian mathematician who had an uncanny ability to just *see* mathematical truths. Then he would report these truths and other mathematicians would provide proofs for them. (Putnam 1975, 64-9) For a moment it seemed to me that all the proof did was to certify in a certain way what was already perfectly well known.

M. The case is very interesting. I have often had the experience in my mathematical work of seeming to see something, and then searching for a proof of it. As if the proof would show me that I had really *seen* what I *seemed* to have seen.

I hope we get to talk about this later. Maybe we could do so in connection with the idea that we can use something like empirical induction in mathematics. But for now I think we should stick with the idea of mathematical *truth*, even if we think of proof as a way of certifying truth rather than as a way of discovering it.

P. Yes. For it now seems to me very important to say that pure mathematics is not merely a formal game and yet does not depend for its sense on having applications.

If so, it *cannot* be a mere artful manipulation of empty signs, but rather must be a use of meaningful signs whose meanings are independent of their applications. And so it *must* deal with some kind of reality of its own.

C. I too feel troubled at the idea that pure mathematics is just a formal game—it really doesn't *feel* like that. But it certainly doesn't concern any reality 'of its own'—whatever that might mean!

M. How does mathematics feel to you. When you do math, doesn't it feel like you're thinking *with* the symbols, not thinking *about* them?

C. Yes. Just as when I think something out in words. I think with the words, not about the words.

M. And so it feels like you're thinking *with* the symbols and *about* something else.

C. Exactly.

M. But then why don't you agree with P that there are the numbers. You could just use the word 'number' as your label for what you're thinking about when, in pure mathematics, you think *with* the mathematical signs.

C. Well—that feels perfectly natural *until* I ask myself what these numbers are. And then I remember the very strong reasons for thinking that they exist neither in space nor in time. And I just can't bring myself to believe *that there are* any such things.

M. So, in a way, it is Frege's answer to the question of what numbers are which leads you to say that there cannot be any.

C. Yes. I suggested that he should not have taken it for granted that there are numbers. I now want to say that he was right to first ask what numbers are, but then, having answered this, he should have asked himself whether there are any. He would have found, in his very conception of number, a reason for thinking that there are none.

A. As you were talking just now it struck me that you are committed to intuitionism.

C. What do you mean?

A. Well, you identify mathematical truth with provability, and intuitionism denies the law of excluded middle. But if mathematical truth amounts to provability, then the law of excluded middle fails, and so must be denied. And this is the intuitionist position.

C. But why say that if mathematical truth amounts to provability, then the law of excluded middle fails?

A. Well, as we've known for quite a while, there are mathematical sentences which are undecidable. Let S be one of those sentences. Then, neither S nor the negation of S is provable. So, you must hold that neither is true, in which case you must hold as well that their disjunction is not true. So, you are committed to the *non-truth* of certain instances of the law of excluded middle. But if you accepted the rules of classical logic, then each instance of that law would be provable and thus, according to you, true. But it would be inconsistent to hold that a disjunction is both true and not true. And so you must reject the law of excluded middle.

It's as simple as that.

C. Couldn't I hold that the disjunction is true though neither of its disjuncts is true?

A. That is a very queer suggestion. The very meaning of 'or' rules out that possibility.

C. Perhaps. But then how about this: We simply drop the term 'true' and speak solely in terms of provability and disprovability.

Then using classical logic I would be committed only to this: that certain disjunctions are provable even though their disjuncts are not.

We certainly are prepared for this outside of mathematics. We accept as provable a disjunction like 'All dodo's chirped or not all dodo's chirped' even though it may be quite impossible to prove either of its disjuncts.

A. Yes, but in this kind of case there is, in addition to what we can do by way of *establishing* the truth of the disjunctions, that in virtue of which they are true or false—the world. And the structure of the disjunction of a sentence with its negation insures that one of the disjuncts will be true in virtue of how the world is, and *that* is what makes the disjunction true.

But in the case at hand—mathematics as you construe it— there is nothing to make for truth *except* provability.

C. I think that you're looking at true disjunctions as 'drawing' their truth from the truth of one or both of their component sentences. So that even in the case of the disjunction of a sentence with its negation the disjunction is true in virtue of how the world is. Either the world makes for the truth of one disjunct, in which case the disjunction is true, or it makes for the truth of the other disjunct, in which case the disjunction is true. But in either case it is how the world is which makes for truth.

A. Exactly.

And so, if there is nothing beyond proof to make for truth, the truth of the disjunction would have to come from the *proof* of at least one of its disjuncts. On the your view of mathematics, there is nothing beyond proof to make for truth. So, on that view there *cannot* be proof for a disjunction except via proof for one or the other of its disjuncts.

C. That *seems* right. Yet I think it may not right.

My basic idea is that nothing except proof can make for mathematical truth. One way of capturing that idea is simply to say that there is *no truth* in mathematics, only proof.

Part of what misleads us here is a certain use of the word 'proof'. We are inclined to say that proof *shows* truth. But I hold that in mathematics proof *is* truth. And that comes to this: truth is proof.

So, if I am right, it is always wrong to say that a proof establishes the truth of a mathematical proposition.

In effect, I now want to say that there is no truth in mathematics, only derivability from axioms.

Suppose now that the *system of derivation* is classical. Then, for each sentence S of mathematics, the disjunction of S with its negation will be derivable from the axioms. Even when neither S not its negation is derivable from the axioms.

Now — what is the problem in *that*?

A. The problem, to repeat it one last time, is this: then your view marks as true a disjunction neither disjunct of which is true.

C. No! My point is precisely that my new view does *not* 'mark as true' any of the propositions of mathematics. In particular, that a proposition has a proof, is derivable from the axioms, is not, for my view, a mark of anything beyond itself.

A. So you are now changing your view. First you held that an arithmetic truth is the same thing as a provable arithmetic sentence. Now you are holding that there is no such thing as arithmetic truth.

C. Yes.

My idea now is to simply *drop* truth and say that there is *no such thing as truth* in mathematics, only proof and disproof.

And this, I think, is what is *basic* in formalism.

One might also put formalism this way: It is the view that in mathematics the most fundamental notion of correctness is not *truth* but *provability*.

In any case, would *this* view (whether or not you call it formalism) need to deny the law of excluded middle?

A. Well—it would have to count as provable disjunctions neither disjunct of which is provable.

C. Yes. But is there a problem in that?

A. Sure. For if the disjunction is *provable* then it can be shown to be true, but it can't *be* true unless it has a true disjunct.

C. Now I really am repeating myself! For I have said that the *formalist* notion of proof is not proof *of* truth. I already cautioned against this misunderstanding. That is why I suggested using the phrase 'derivable from the axioms'.

So now I will ask: Is there a problem in having disjunctions which are derivable from the axioms even though their disjuncts are not derivable from the axioms?

A. I'm not sure. I sense there must be a problem here, but I don't think I actually see one.

C. Look–the *actual* situation in classical mathematics is of just this kind. For each mathematical sentence S, the disjunction of S with its negation *is* derivable from the axioms, *and* unlimitedly many such sentences–the one's which are the disjuncts of these derivable disjunctions–are *not* derivable from the axioms.

So unless you see a problem in classical mathematics, you won't see a problem here.

A. That feels convincing.

C. As I now tend to see the matter, formalism must be *neutral* between classical and intuitionist logic. Each is simply the calculus it is.

A. But might there not be a problem in classical mathematics–as the intuitionists allege?

C. Perhaps, but this is not something to which formalism can speak one way or the other.

For formalism says that the language of pure mathematics is a notation of meaningless marks combining into sentences devoid of sense. The language has the *formal* structure of a standard kind. But it is all form without content. We of course *could* give meanings to its marks and senses to sentences, and perhaps in a number of different ways. But in pure mathematics, according to the formalist conception, we *don't*. For the rest it is a matter of producing certain patterns of sentences in accordance with logical rules: derivation.

A. I think I see the force of those remarks. But if they are right, then formalism is not much of a philosophy of mathematics. So far as formalism goes *any* notation manipulated in accord with *whatever* rules would as much be a pure mathematics as the one we actually employ.

C. I agree that formalism omits much of great importance. But for all that, its fundamental claim that the most fundamental conception of correctness applicable to the sentences of pure mathematics is that of provability in a sense which is not the provability *of* truth, or *of* anything, may yet turn out to have been a *fundamental* insight.

But these are enormously difficult matters on which to gain any real clarity, and I think we should only conclude that we *might* have come to see that formalism is independent of intuitionism through this discussion.

CHAPTER 6
A Conversation on Existence and Number

In this dialogue there are three characters C, P, and M. In the last dialogue P took a Platonist position. In the beginning of this dialogue P is challenged by C, who takes an anti-platonist line. M is not so easily classified, taking a rather skeptical line with respect to both Platonism and anti-platonism. Early in the dialogue the existence of numbers is taken up by P and M. M defends a view similar to that put forth by Rudolf Carnap many years ago (Carnap1978, p. 153). The main questions are: whether 'There are numbers', with its philosophical import, is provable in arithmetic; and whether it is even expressible in arithmetic.

The Dialogue

C. Of course there aren't any numbers. They aren't here or there or anywhere. What's nowhere doesn't exist at all.

P. You say that numbers don't exist because they're nowhere. But how did you find out that what exist has got to be somewhere? You're like a man looking into the wrong end of a telescope. It is *plain* that there are numbers. And so, since its also plain that numbers lack location (and duration as well, I might add), you can conclude that there is no need for something to be somewhere for it to exist.

C. That only sounds good. But what's really obvious is that you'd better have some pretty good reasons for saying that something exists even though its nowhere. You tell me there are numbers. I ask you where they are. You reply that they're nowhere. Strange entities these numbers! Tell me how you found out that there are these nowhere things!

P. I "found out" about numbers like anyone else. I learned to recite
the numerals in order, and then learned to count, and then
learned to add and multiply. Pretty soon I had the hang of it.

C. So you operated with words in various ways. That's all you're
telling me so far. I want to know: How did you find out that there
are numbers—that there is something more to what you learned
than your learning to use the numerals in counting, adding,
multiplying and the like.

P. Like anyone else, when I added some numbers I found the
number that was their sum. I was finding things out about
numbers. In doing this I worked with numerals. But I didn't find
things out about *them*.

C. So you say. But it seems strange to me that you could find
something out about numbers just by fiddling with numerals. It
would be like saying that you could find something out about
people by fiddling with their names. You operate with numerals
and never look beyond them to anything else. So this talk about
numbers is just more numeral talk.

P. When I calculate I'm not just fiddling with numerals. I operate
with the numerals and thereby discover certain sums. I certainly
don't find out that '2' and '3' add up to '5'. I don't even know
what that *means*—to "add up" numerals. And neither do you.

C. I'm not saying that operating with numerals enables us to find out
that *they* add up to this or that. The whole manner of speech of
"sums" and "products" is carried out in the *use* of numerals. It's
more *numeral talk*. It is talk in terms of the term 'number' as well. I
say that the *number* that is the sum of 3 and 2 is 5. But the word
'number' can be dropped, along with the word 'sum'. It is enough
to write '5 = 3 + 2'

P. I agree that 'number' can go. But 'sum' just gets replaced by '+'
and we continue to *use* the numerals, not talk about them. What
you fail to see, but surely will see now that I point it out, is that
we use the numerals, the addition sign, the identity sign and the
rest to *say things*—for example, that 5 equals 3 plus 2.

I agree also that the fundamental question in the philosophy of mathematics is whether or not there are numbers. *That* is the question that divides philosophers of mathematics.

M. Well, philosophers produce such sentences as 'There are numbers' and 'There aren't any numbers', and take themselves to thereby be making assertions on which they differ—one asserting what the other denies.

But is it *obvious* that there is either assertion or denial in such exchanges? Doubtlessly it *feels* like assertion and denial.

P. Surely it is possible to assert that there are numbers, to do more than produce the sentence with the feeling of assertion!

No one doubts that we assert something in the use of such mathematical sentences as

> For every number *n* there is some number
> *m* such that m is prime and greater than *n*

and so, since

> There are numbers

is a logical consequence of this mathematically correct sentence, we also assert something in its use. Surely the logical consequences of what can be asserted can also be asserted.

M. That actually isn't clear to me. But I want to bring up another point. Take a look at

> For every number *n* there is some number
> *m* such that *m* is prime and greater than *n*.

Is 'There are numbers' a logical consequence of this sentence?

In the mathematical sentence the word 'number' is part of the *quantifier* phrase 'for every number *n*'.

P. Are you suggesting that 'number' is a quantifier?

M. No—not that. I am just pointing out that in the mathematical sentence the word occurs within a quantifier phrase. And otherwise it isn't in the sentence at all, and so it isn't in the sentence as a *predicate*, and so there isn't the logical consequence you thought there was.

P. That's just logic chopping. But set aside my little argument involving consequence. The sentence 'There are numbers' is itself a perfectly good mathematical sentence, and so it makes sense to assert it.

M. Isn't it even the case that 'There are numbers' is not a sentence of mathematics?

After all, 'There are numbers' contains no *mathematical* signs.

P. What do you mean? It contains the word 'number'!

M. But is that a mathematical sign? After all, we can do quite well without this word in mathematics, and actually *use* no such sign in, for example, what we call 'natural number theory'. In developing this theory we use such signs as '+', '·', '=' and the numerals, along with such letters as '*n*' and '*m*' together with signs for generality and sentential composition.

The *mathematical* signs are the operators, along with the numerals and letters.

Among these is *not* to be found a sign corresponding to the word 'number' in the kind of grammatical application it is given in 'There are numbers'.

P. How can you say that the word 'number' is not a mathematical sign? It is right there in the sentence about primes!

M. The *mathematical* sign which is right there in the sentence about primes is the word together with its use—and, as I pointed out earlier, its use in that sentence is as a part of a numerical quantifier. *That* sign is absent from

There are numbers.

And apart from this–its use in expressing mathematical generality—the word plays no role in arithmetic.

P. I now see what you mean. But this is just an accidental feature of our arithmetical notation. It shows nothing of importance. After all, we could *introduce* into the language of mathematics a sign of the kind that occurs in the philosopher's sentence, and then it would be a sign with mathematical sense after all.

For example, it would be enough to actually form the practice of constructing sentences in which 'number' occurs in the sorts of positions accessible to such defined mathematical terms as 'prime', and then accept every sentence of the form

n is a number

for numeral n.

M. So your idea is that by means of this procedure we will *bring it about* that a particular sign has a mathematical use, and thus a mathematical sense. In that case, the sign *as yet* lacks such use or sense. So let's use a less suggestive word as our new sign. The word 'zib', for example.

In conformity with the procedure sketched above I now construct the sentence

3 is a zib.

Have I thus far said or asserted anything? Has my construction endowed that word with any sense at all? No one supposes it has, for I have so far done nothing beyond mouthing the words '3 is a zib'.

Suppose I decide to introduce into English the word 'tsw' and express a readiness to accept every sentence of the form

p is a tsw

for proper noun p. Now—what sense have I thereby attached to the word? None at all.

Now consider the rest of the procedure—the part described as 'accepting every sentence of the form … '.

In what is this 'accepting' to consist?

Would it be a matter of declaring each such sentence *true*?

But is it enough to murmur 'True' at a sentence for that sentence to be true? Mustn't a sentence *already* have a sense for it to be true? How can *calling* it true invest it with that which it needs to *be* true?

Or does 'acceptance' here come to this: When someone writes, e.g., '13 is a zib' I do not raise a fuss. I let it pass. But if someone writes '13 is not a zib' then I raise a fuss about it. But *what* will I raise a fuss about?

Suppose I hear you utter the words '13 is not a zib'. I then hoot and holler. Someone may ask me, Why the fuss? How should I respond? What bothers me?

Can I honestly say that I am raising a fuss because he said something *false*?

I might in fact fuss at such an utterance—by pointing out that it has not as yet been given any sense!

P. That is an almost willful misunderstanding of what I meant. Of course it is not just a matter of making *sounds*. That obviously is not sufficient for sense. The important thing is not that we make certain sounds but that we *affirm*, e.g., '13 is a zib' but *deny* '12 is not a zib' and reject it in favor of its negation. That is what gives the word a use and thereby sense.

M. Part of this was clear to me from the start. I know I am to *say*, to *utter the words*, 'No—12 *is* a zib' when someone utters the words '12 is not a zib'. But have I thereby *denied* anything? Or *affirmed* anything?

As yet the words have no meaning.

To deny something with the words 'No—12 *is* a zib' the sentence '12 is a zib' must already have a sense. So, the very procedure that was to fix a sense for sentences formed with 'zib' presupposes that a sense has already been fixed for such sentences.

I can easily master the suggested routine with words, and even make it part of my practice with words. Others may go along with it. But it doesn't come to anything.

If we eventually drift away from this use of the word, change it a bit here or there, even suddenly, who cares?

It really doesn't matter what sentences we construct using 'zib'—for by such constructions the word has not been put to work.

P. Well then, we can supply a sense for the word by introducing it into arithmetic via expressions which no one doubts to possess a mathematical sense. It will be enough if 'zib' has the sense of the arithmetical formula which matches it in its intended extension, and to this end it will be enough to use this word as short for that formula.

M. How can 'zib' have the sense of *the* arithmetical formula which matches it in its intended extension? After all, for each mathematical formula there are unlimitedly many more arithmetical formulas which match it in extension—just as '$n = 2$' is matched in extension by '$n = (3-1)$' and '$n = (1+1)$'.

So this too will fail to fix a sense for the word.

Or should we say that it is to have *the* sense of *all* such formulas? But they have *different* senses.

Shall we say that it is to have *every* sense of each such formula? Then it will be indefinitely ambiguous.

I asked for what shows that 'There are numbers' makes sense. The idea was that this could be shown by showing that it is a sentence of mathematics—for we don't doubt that mathematical sentences make sense.

But then we noticed that 'There are numbers' is *not* a sentence of mathematics, and, further, that adding 'number' in the manner suggested fails to secure a sense for that term, and thus also fails to secure a sense for the sentence in question.

P. But why worry about the specific *form of words* 'There are numbers'. Just select some one overtly mathematical sentence, for example

There is an n such that for some m, $m = n + 1$,

which contains a formula true of exactly the natural numbers, and use *that* formula to formulate the dispute at hand.

M. But what makes 'for some m, $m = n + 1$' a formula of natural number theory in the first place? Surely it is essential that the letters 'm' and 'n' be variables for natural numbers.

And what makes them that? Well, it is our practice to admit only numerals and numerical expressions formed therefrom as replacements for such letters in inference.

As to what marks an expression as a numeral, that will be its role within arithmetic *together with* its applications in empirical statements of number.

P. Perhaps. But how do these considerations constitute an *objection* to our selecting just any numerical formula which is correct no matter what numeral we might construct and put in for '*n*', for example

$$\text{for some } m,\ m = n + 1$$

as our surrogate for 'number' as used in philosophy?

In fact they are no objection at all. So let's go ahead and conduct the philosophical dispute in terms of *this* unexceptionable mathematical formula.

M. But if '*n* is a number' is to be taken as short for 'for some *m*, *m* = *n* + 1', then the philosopher's

$$\text{There are numbers}$$

comes to (has the sense of)

$$\text{For some } n \text{ and } m,\ m = n + 1$$

in *its* mathematical sense. And its having that sense depends on it being our practice to admit only numerical terms as replacements for the letters. And this is not expressed by the mathematical formula, since it is a condition of the formula being mathematical. That condition is, of course, fulfilled. Given that, the formula has a quite particular *mathematical* sense. It is just *another piece of mathematics*.

If some philosophers find that all they wish to say is that there is an *n* such that for some *m*, *m* = *n* + 1 (that, as we might put it informally, there is a natural number which yields a natural number by addition of one), then they will have found that they wish to say only what virtually everyone agrees to—something entirely *mathematical*.

Ask yourself whether the *entire* content of mathematical realism is *entirely* mathematical in nature. So that the propositions of mathematical *realism* are nothing other than theorems of arithmetic.

This picture of a person who from time to time—and perhaps in books or articles–writes down various well-known arithmetical quantifications and calls it a *philosophy* of mathematics fits nothing I am familiar with.

Someone says 'There are no numbers'. The Zen master replies with a theorem of arithmetic, and perhaps lays out its proof. He meets every such challenge in just this way. He might also, from time to time, apply the conclusions of certain proofs in some practical way. I might call him a philosopher, but I would not say that *what* he says constitutes a philosophy of mathematics.

In any case, the mathematical realist is no Zen master. The realist wants to *add* something—namely, that these theorems and proofs are not about *nothing*—that *there are* things (that there are *things*) which they are about—*numbers*.

But to add this is to utter a sentence *in addition to* those of mathematics, and so to utter a sentence *not* a part of mathematics, and thus *not* a sentence the sense of which is secured by its place within mathematics.

It is in going beyond saying what makes mathematical sense (something internal to mathematics) that the realist differs from the Zen master.

Realists may present various proofs within mathematics and make various assertions within mathematics. But then they want to add something, and try to do so with such words as 'There are numbers'. This is something they can't so much as *try* to do if they stick with the sentences of mathematics.

C. I think I am getting this. You are making this point: we cannot show that the realists' sentence makes sense by noting that it is just another mathematical sentence for which it is unproblematic that it makes sense.

M. *And what holds for the realist holds as well for the anti-realist who disputes realism by negation!*

If 'number' is short for some *mathematical* formula, then 'There are numbers' is just another quite ordinary mathematical sentence, one of a kind which goes virtually undisputed. And if 'number' is *not* short for some mathematical formula, then one certainly cannot show that 'There are numbers' has a sense by showing that it is a sentence of mathematics!

P. How about this then? In logic, when we want to say something like 'Socrates exists' we say 'For some x, x = Socrates'. And so to say that something exists we say 'For some x and y, $x = y$'. And so, 'Numbers exist' gets expressed by 'For some n and m, $n = m$'. That is a perfectly good sentence of mathematics, and so there is no bar to asserting it.

M. But doesn't 'For some x, x = Socrates' say that Socrates is identical with something?

   We have the English sentence 'Something exists' and cannot paraphrase it into the notation of our logic because that notation handles 'exists' quantificationally. So we *can't*. And paraphrasing it by 'Something is identical with itself ' is just something people agree to so as to get past a problem.

   It even is *obvious* that 'Something exists' and 'Something is identical with something' are not equivalent sentences.

   No, if you want to capture 'Something exists' in our standard logical notation you'll need just what you find in *English*—a predicate. Let it be 'exists'. We can then write 'For some x, x exists' as our paraphrase for 'Something exists' and 'For some n, n exists' as our paraphrase for 'Numbers exist'.

   And then it will come as no surprise to you when I point out that 'exists' in its use as a predicate is a *not* a mathematical sign. It is, of course, a sign which occurs in a quantifier phrase which has a perfectly good use in mathematics, namely 'There exists a number such that', but that occurrence is not as a predicate!

P. Look—among the perfectly good sentences of mathematics is 'There are prime numbers'. So let's just formulate the dispute using *that* sentence.

M. What dispute? Who disputes that there are prime numbers? Is there anyone who thinks that no numbers are prime?

P. Of course no one thinks that there are numbers but that none are prime. But some people nonetheless think that no numbers are prime—because they think there are no numbers.

M. So the dispute is not about primes. Rather, it is a dispute about whether there are numbers. This brings us right back to where we started—the *non*-mathematical sentence 'There are numbers'!

P. But the sentence 'There are primes' can formulate what is in dispute, for it is a *quantification*. And so it asserts the existence of numbers.

M. Yes, 'There are primes' is a quantification and says that there are primes—it asserts the existence of primes.

   Virtually no one denies this point.

P. But haven't you denied that very point again and again, by implication if not in so many words? Hasn't your underlying idea been that no mathematical sentence asserts the existence of numbers?

   That this idea is in error is shown by the fact that, for example, the sentence at hand is both mathematical and true, and, since it is an existential quantification, true only if there are numbers.

   The mathematical realist merely affirms what is already asserted within mathematics by its existential quantifications—that there are numbers!

M. Of course the sentence says that there are primes. The sentence says what it says and says that in so many words. But how can its saying that there are primes make it clear that anything is said with the sentence 'There are numbers'?

   The quantification *says what it says*. You say that it says that there are numbers. I say that it says there are primes.

   And I question whether you *do* say (do *say*) that it says that there are numbers. For I question whether 'There are numbers' says anything. And if it doesn't, then the same holds for 'The sentence says that there are numbers'.

   What I of course do *not* question is that you use the words 'There are numbers' when you seek to say some of what is said by 'There are primes'.

P. The sentence 'There are prime numbers' indeed says what it says, and what it says is that there are prime *numbers*. And so it says that there are numbers.

M. Is the sentence a sentence of mathematics?

P. Yes.

M. So what it says is what gets said in mathematics. But if I recall rightly 'There are numbers' is not a sentence of mathematics. So how can a sentence of mathematics say what goes unsaid in mathematics?

P. This whole involuted enquiry is predicated on the supposition that there is a doubt whether 'There are numbers' makes sense. But there is *no* plausibility in this so-called doubt. The words at issue are plain English. You may *claim* not to understand the sentence 'There are numbers', but you understand it nonetheless. This doubt is just a pretense, and the enquiry to which it has led has no point.

M. The sentence 'There are numbers' has a normal grammar and its words are familiar ones. Does that show that it has a sense? If it did, it would show as well that the sentence 'Three is red' has a sense. And it isn't *obvious* that it has a sense.

I grant that it *may* be that we *do* perfectly well understand this sentence, that it does make sense and that we all grasp the sense it makes. It *may* be that it is only a false philosophy that keeps me from seeing clearly that this sentence makes a sense I grasp. But— for whatever reason—it yet is not *clear* to me that it says anything. In *that* sense my doubt is a real one, even if it is founded on nothing but the obscurities in my own thought.

P. You need to relax and learn to accept the obvious. It may help if you look at the matter this way: We all grasp the concept of number, and so surely can conceive that something falls under it. This much is accessible to you, for you have not gone so far as to call into question our having the *concept* of number!

M. There is no doubt that we grasp the concept of number. But what kind of concept is it? Is it a concept of a kind such that there are concepts of that kind under which things fall?

Not every concept is a concept of that kind. The concept of existence (as expressed in a sentence like 'There are lions') is not a concept of that kind, as is shown by the fact that what expresses this concept is a *sign for generality* (e.g., a quantifier), not a general term (e.g., a noun).

P. So you really are saying that 'number' is a quantifier.

M. No. I used the example of the concept of existence to show that not *all* concepts are predicative in nature. When I say that the concept of number may be like the concept of existence I mean only that it may be like it in *not* being predicative.

Nor am I suggesting that the concept of number is *not* expressed by our arithmetical language. But *what* about that language expresses the concept of number?

Might it not be the *letters* '*m*', '*n*' and the like?

We construct formulas using these letters and replace these letters by numerals and carry out certain inferences with formulas using these letters along with formulas resulting from these by replacing letters by numerals.

Won't it be (roughly) our grasp of all this which constitutes our grasp of the concept number — not our mastery of one or another mathematical *predicate*. (And what we would need for 'There are numbers' is the kind of concept expressed by a predicate.)

Remember: Even if we had 'number' as a predicate in the language of natural number theory, it would be entirely useless.

Our *actual* use of 'number' is for the expression of generality. It's use is like that of a subscript which reminds us which expressions are to go in for the letters.

P. But there is nothing essential in this. After all, as Quine has repeatedly pointed out we can always dispense with variables of different sorts in favor of variables of a single sort. All we need do is introduce predicates to do the sorting. (Quine 1960, p. 160, pp. 229-232)

So, instead of saying

For some number *n*, *n* is prime

we can say

> For some *n*, *n* is a number and is prime.

M. Quine encourages us to rewrite the old sign for generality 'for some number *n*' by a *new* sign for generality 'for some *n*, *n* is a number and'.

By means of this shift in the form of expression nothing of mathematical consequence is brought about. 'Number' is now in a grammatically predicative position, but does *no mathematical work* of the kind appropriate to expressions accessible to such positions.

It all goes back to the point that arithmetic actually lacks any such term as 'number' except as a part of its signs for generality. And there too it is dispensable, and is actually dispensed with when we get down to formal proofs.

For formal proofs make do with just with the letters. What 'does the work' of the word 'number' in its occurrences in signs for generality is our *practice* of replacing those signs just by numerals and numerical terms formed from numerals.

It is *that* practice which makes 'number' mean *number*!

If it were our practice to replace '*n*' in those of its occurrences within the scope of 'for some number n' by names of people, 'number' would not mean *number*.

It now *seems* that what expresses the concept *number* neither is nor could be any particular mathematical sign, but rather must be something about how we operate our system of arithmetic—how we calculate within it, and how we apply it.

We instantiate to *numerals* within arithmetic and *use* numerals in such and such ways in the sentences exterior to arithmetic.

It is *this* aspect—this *complex* aspect—of the language of arithmetic that expresses the concept *number*. And *this* cannot, as it were, be pulled back up into arithmetic and made the content of one of its signs.

The concept *number* is involved in our thinking *about* a certain conceptual practice, not in our thinking *within* that practice.

P. In that case it is of no significance whatsoever that we can do mathematics without the predicate 'number'. It will be enough that we actually use that word predicatively, even if only outside of mathematics. And you have just agreed that the concept number is available outside of arithmetic, that 'number' has its extra-mathematical uses.

And that *is* how we use it when, for example, we distinguish between colors and numbers. We say that three is number, but that red isn't.

Or just imagine the use of mathematics in a physical theory. We there can use 'number' in a predicative manner so as to distinguish, e.g., particles from numbers.

It is like the case with sets. *Pure* set theory does without any sign for sets—but only because in pure set theory our domain consists of classes alone. But the domains of those languages in which set theory finds an application are not thus limited—and within them a predicative sign for sets finds a use.

M. So I now need to consider such sentences as 'Colors are not numbers', 'Red is not a number', 'Particles aren't numbers', and 'Tables aren't sets'.

Suppose that the following is a truth formulable in the language of some theory of color including at least elementary arithmetic:

For every x, if x is odd, x is not even

We will not want to infer, e.g.,

If red is odd, then red is not even,

and to bar this inference we could use the word 'number' and write instead

For every x, if x is a number then if x is odd, x is not even.

But if we do, we can then infer

If red is a number then if red is odd, then red is not even

which is equally unwanted. What we really need is

For every number x, if x is odd, then x is not even

which yields

> If 3 is odd, then 3 is not even

but not

> If red is odd, then red is not even.

So, what is needed to guard against nonsense is not a *predicate* 'number', but the use of that word to delimit acceptable substitutions for letters used to express numerical generalities. And what is essential here is the *practice* of replacing certain letters *only* by numerical terms. And so once again we do not need the word as predicate to draw a distinction among things.

The fundamental point would seem to be that the language games for color and number don't 'intersect'. A term with the use of a color word won't in general yield a sense when it replaces a term with the use of a number word, and conversely. Neither 'Red is the product of 3 and 2' nor '3 is brighter than pink' make sense. We can of course count each *not true*, and in that sense false. But in *that* sense a can opener also is false.

Just as 'number' serves as an index to generalization, and thus is dispensable, so also for 'color'. Suppose we lacked this word. We yet might say that Joseph had a coat of many colors by using a sentence like

> For many f, Joseph's coat was f

where it was our practice to recognize as instances of this generalization only such sentences as

> Joseph's coat was red
>
> Joseph's coat was blue

and the like—that is, as *we* would say, to recognize as instances of this generalization only sentences formed with words for colors— words with *that* kind of use.

P. You asked for something which supports the claim that 'There are numbers' makes sense. But why? Could *you* produce something which supports the claim that 'There are beets' makes sense? And if you couldn't, would that at all sap your confidence that it does?

That we are not sure how to *show* that a sentence makes sense is *no reason whatsoever* for doubting that it *does* make sense.

M. Consider the case of beets. Someone might read a description using 'beet', and then ask to be shown that there are such roots. We know how to respond to this request. We bring various roots and see whether any fits the description, and find one does. We then agree: Yes, there are beets.

There is something analogous for 'square of 27'. An easy calculation shows that 729 is a square of 27. Having carried it out, we will then agree that Yes, there are squares of 27.

But the case for *number* doesn't fit this familiar pattern.

What description do we have for 'number', so that we can decide whether 37 fits that description? Shall we say that a number is a timeless entity? By what method might we find out that 37 fits that description? Or, what calculation shows that 37 is a number? No *calculation* shows any such thing.

A *calculation* leads to '37 = 23 + 4 + 10'. Nothing in mathematics leads to '37 is a number'—that is not something which can be brought out *within* mathematics.

Someone unfamiliar with our notation might ask whether 37 is a number. We could then exhibit our use of that sign. They would then be satisfied. '37 is a number' can be used to express a recognition about the *use* of a sign.

And in finding out that 37 is a number, one will not find out that alone. For the numerals form a system. A person could no more have only one number word than a person could have only one color word.

We master a technique in both cases. In learning to count we learn a method for syntactically arriving at each of *many* signs, and for *using* them as numerals to make statements of number (e.g., 'Lela has 2 dogs'). The number words do not come 'one by one'. We possess a repeatable technique for constructing empirical sentences. And something similar holds for color words.

P. That seems wrong on both points. A child can recognize a color wordlessly, and then be taught a word for just that *one* color. And there are societies in which counting goes to three and no further.

M. Yes, a child might come to utter 'red' only when something looked red to him. But would that make 'red' *in his use of it* a color word? Would he have been taught a 'word for a color' in being brought to utter 'red' only when something looked red to him?

   This use of 'red' would be a tiny fragment of our use of 'red'. And so there would be a certain similarity. But also vast dissimilarity.

   My *guess* is that on this use of 'red' there is an important connection between the word and the color (seeing the color prompts the word given certain other conditions), but (here is the guess) not one of a kind which makes the word a color word.

   An animal with a distinctive response to red does not therein have a color word.

   Similarly, a society may possess a tiny fragment of our use of the numerals. But does this make *their* '3' a number word? Again there will be a bit of similarity, along with vast dissimilarity.

P. It goes from bad to worse. First you doubted whether 'There are numbers' makes sense, and now you've denied that three is a number. That is absurd. It is like denying that three is odd, or prime.

M. That three is a number has not been denied. Rather, it has been asked what sense, if any, is made by the sentence 'Three is a number'. The sentence 'Three is prime' belongs to mathematics. But the sentence 'Three is a number' is no more a sentence of mathematics than is 'There are numbers'.

   You might suggest that we take '3 is a number' to have the sense of the sentence

$$\text{For some } n,\, n = 3?$$

But this sentence is a logical consequence of '3 = 3' and so says no more than *it* says. And does the identity say of *anything* that it is a number?

Or should we take '3 is a number' to have the sense of the sentence

For some n, n is a *number* and $n = 3$?

But now we must ask about *this* occurrence of the word 'number'. What does 'n is a number' come to? Presumably this:

For some m, m is a number and $m = n$

in which case we have

For some n, for some m, m is a number and $m = n$ and $n = 3$

at which point we need to work on 'm is a number'!

'3 is a number' is not the result of any proof or calculation. In fact, the sentence is *arithmetically* useless.

So, whatever sense it is made to bear is one it will obtain from some use *external* to arithmetic.

What needs to be resisted here is the idea that we gain *further insight* into numbers when we step outside of mathematics. What we obtain *outside* of mathematics is not continuous with what we obtain by carrying out proofs and calculations.

It isn't as if we gained insight into the objects of mathematics first by working within mathematics, and then, by stepping outside of mathematics, gained some additional insight into those objects. By stepping outside of mathematics we do gain *something* — but what?

P.   This is just so much palaver. The key point gets *lost* in all this talk. It is *clear* that we *all* realize that three is a number and that red, for example, isn't. And this recognition can *easily* be expressed in words — as easily as it *has* just been expressed in words!

M.   If there *is* a recognition here, then it might be lacking. So let us suppose that someone failed to recognize that three is a number. What would they have missed? And how might their failure be remedied?

I here imagine a person who counts, adds, multiplies, applies the results of adding and multiplying, etc. I imagine a person reasonably   competent   in   the   empirical   application   of mathematical terms, and in the arithmetic of those terms. This is enough to have him be one who grasps the idea of three. But he is supposed to *fail to recognize* that three is a number.

So now we tell him something about numbers. Our hope is that once he gets this information he'll recognize that three is a number.

What do we tell him? Shall we say that a number is a timeless, placeless entity?

The doubt whether anything is asserted by 'There are numbers' is matched by the doubt whether anything is asserted by 'Three is timeless' and the like.

(But *must* recognition be coupled with the *possibility* of a failure to recognize? Aren't there certain facts that *can't* go unrecognized once they are considered? If I can *ask* myself whether a point in visual space could be both red and blue, then I *cannot* fail to see that it couldn't.)

P.   Look, we say such things as that such and such is a number of a certain kind, e.g., prime, or negative, or irrational. This is a matter of distinguishing different kinds of numbers—just as we might distinguish different kinds of birds.

In a math class the teacher writes down the word 'Numbers' and then subdivides—just as a teacher might do in a biology class, beginning with the word 'Birds'. And we do speak of *kinds* of numbers.

To say there are numbers is, then, to say the sort of thing said by saying that there are birds. The sentence 'There are numbers' makes *that* kind of sense.

M.   'There are birds' is like 'There are primes'. 'There are numbers' is more like 'There are material objects'.

P.   But surely the sentence 'There are material objects' makes sense— it even is a sentence which *could* be false.

M.   That may be more easily said than thought. Ask yourself what would lead you to seriously suppose that there are no material objects? You perhaps imagine your body turning into a vapor and the same for everything else in your environment. A world of wisps. But wisps are material objects.

What are we to imagine? An immaterial eye eyeing nothing at all?

P. *Really*! As if it were *impossible* for there to be no material objects! And what difference does it make whether or not I can *imagine* it? It is enough that I can *think* it.

M. Yes—it is in a way possible for there to be no material objects. What sort of possibility is that? If it is only *logical* possibility, then remember it is logically possible that three is even.

There are sentences grammatically like 'There are material objects' which are false. That is, the sentence is not logically true. So, it is not *logically* impossible that there are no material objects. But that does not show that it is thinkable that there are no material objects—or possible, except perhaps in the way that it is possible, e.g., that 3 is even. And this example shows that not everything that is logically possible is thinkable.

I *suspect* that we think nothing at all with the words 'There are no material objects' *or* with the words 'There are material objects'.

P. Suspicions indeed! How are we to reply to that? But leave it aside and note that we can and do speak of different types of numbers. That alone is enough to show that 'number' makes sense as a predicate.

M. Primes are one kind of number. Evens are another. The distinction gets made within mathematics in familiar ways. And how it gets made within mathematics does not draw on any cooked up predicate like 'number'.

The concept number is not available *within* arithmetic—it is not a concept with an arithmetical construction.

A sentence like '3 is a number' is non-arithmetical. It is not a piece of mathematics.

P. Saying it three times doesn't make it true! But it isn't important even if it is true. For don't we also speak about the numbers from a point of view external to mathematics? Philosophers do this all the time.

M. The philosopher's point of view *is* external to mathematics. And for that very reason the philosopher has nothing whatsoever to say about numbers.

Still, it again and again seems that there is something which the mathematician leaves out—something which *mathematics* leaves out—but on which the philosopher can pronounce, for example by saying that there really are such things as numbers, that they really exist (or that there really are no such things).

The philosopher has something *unmathematical* to say about the numbers—for example, that three is one of them and that they exist … or that they do not exist.

If we say they do exist, then we need to answer certain objections. If we say they do not exist, then we need to answer certain different objections.

The 'space' of the philosophy of mathematics seems to be defined by these two options—but is not.

REFERENCES

Carnap, R. (1978). Empiricism, Semantics, and Ontology. In: I. M. Copi and J. Gould eds. *Contemporary Philosophical Logic*. New York: St. Marten's Press.

Gödel, K. (1964). What is Cantor's Continuum Problem? In: Benacerraf, P. and H. Putnam eds. *Philosophy of Mathematics*. First Edition. Englewood Cliffs, N. J. : Prentice Hall.

Frege, G. (1960). *The Foundations of Arithmetic*. Second edition. Translated by J. L. Austin. New York: Harper & Brothers.

Mill, J. S. (1974). *A System of Logic, Ratiocinative and Inductive: Being a Connected View of the Principles of Evidence and the Methods of Scientific Investigation*. Edited by J. M. Robson. Introduced by R. F. McRae. Toronto: Toronto University Press.

Plato (1985). *Meno*. Edited with translation and notes by R. W. Sharpes. Warminister, Wittshire, England: Ars & Phillips Publishers.

Putnam, H. (1979). *Mathematics, Matter, and Method*. Cambridge and New York: Cambridge University Press.

Quine, W. V. (1960). *Word and Object*. London and New York: The Technology Press of the Massachusetts Institute of Technology and John Wiley & Sons.

INDEX

FSC
www.fsc.org
MIX
Papier | Fördert
gute Waldnutzung
FSC® C083411

Zeitfracht Medien GmbH
Ferdinand-Jühlke-Straße 7
99095 Erfurt, Deutschland
produktsicherheit@kolibri360.de